arthous

arthouse

schwartz/silver architects

conceived and edited by joseph giovannini
photographs by alan karchmer

BUILDING MONOGRAPH

designed by linda prescott
directed by oscar riera ojeda

contents

"I believe that the house belongs to everyone who has been part of it. Like my dog. I own it only to the degree I participate in its life. You don't own a thing if it's alive. That's what it means to be a parent, and this house is the same way."

— Eileen McDonagh, owner

"Everyone had in mind how to add to the artistic merit of the house. If you look at the house as a piece of sculpture, the interior furniture is meant to complement, not distract, from it. Then there's the secondary sculpture—the fireplaces, the kitchen—and our pieces of Inuit sculpture. It's really the patron of the arts sort of thing."

— Bob Davoli, owner

foreword

A General Theory of Architectural Relativity

This book borrows from its architectural subject, a house with multiple authors and a moving center. No single hand ruled the collaborative project; no axis gives the Boston residence its orientation; no grid ordains a controlling understructure. The moving center roams the interior and the grounds, offering multiple viewpoints onto the bucolic site's pond and forest. Like the house, the book tells no one story but, rather, the many stories that have gone into its making and interpretation—recounted independently by the owners, architects, artists, contractors, craftsmen, inspectors, and journalists. Inspired by Japanese director Akira Kurosawa's Rashomon, a film in which the same tale is told by several actors, none dominant, this is a story—and a house—without omniscience. The absence of one author shaping the narrative liberates the tale from a single storyline and evaluation. Releasing the book, and the house, from fixed positions allows multiple viewpoints to float in a general relativity of meaning. An image of the whole gathers, like a Pointillist painting, through the gradual accumulation of the many points.

"An image of the whole gathers..."
– Joseph Giovannini, architecture critic

robert davoli **Client, Venture Capitalist, Musician**

I'll tell you: I would rather have been a musician, or a writer. I'm in finance. My background is I founded some software programs and sold them. I'm not really a financial guy. Our whole orientation is different. Eileen and I have no friends who are rich. All our friends, all, are artists or intellectuals. I notice that wealthy people, their minds go numb, they try to anesthetize themselves with 1,500 rounds of golf. They use their money to pamper themselves, and it turns their minds to jelly. I play the guitar. I do this architecture project. I can't imagine retiring in the conventional sense, kicking this little white ball around. We don't have one friend with wealth. You go out to dinner, and realize they're really boring. When they get money, they build a huge colonial or Georgian, and just make it bigger, no creativity. All their wives are into historical. It's not a value judgment, not right or wrong, just different. Most of these guys are straight, just into sports, not fine arts. They think I'm a weirdo. The way they describe the house to friends is that it's a "one-bedroom."

The emanation of the idea for the house came when we started looking for architects. The germ was: Let's find some collaborative architect, like an extemp jazz musician. Not the classicist architect. We'd give the guy a riff, and then he'd riff back. With Frank Gehry, he'd make the house look like a Frank Gehry. We paged through books, and we quickly decided we didn't want a Gehry or a Richard Meier. We wanted an architect whose work didn't look the same from project to project. Maybe we were illogical, but if the work differed, we thought, then the architect must be listening. That was important to us: collaboration. We wanted to make this into a Bach fugue, with your voice, our voice, a design that synthesized many voices.

You know what a venture capitalist does? He makes money by taking a stand. We invest in early-stage companies, tech companies, and they get bought for a good premium or they go public. These endeavors are for profit. But the reward of the house was not financial. It was aesthetic. This venture was not for profit

As we got into the project, we thought, okay, we need a landscape architect and an interior architect, and then it dawned on us that it could really be an art project. It dovetailed into the idea of a piece of sculpture you could live in, or a museum you could live in, so in the combination of house and sculpture, we thought: Why don't we get other artists to contribute? Let's hire some sculptors, painters, and architects for the interiors, and then we'd have a photographer from day one, an art photographer, and she can have a show on this house in a book. Our son's a composer, Robert Price. Let's hire him to write a piece for the house. And the pizza oven guy from Passim's—why doesn't he become part of the collaborative?

As our constituents, there's an inventor, three sculptors, two painters, a composer, the architects, and us. There're thirteen or fourteen of us. Everyone had in mind how to add to the artistic merit of the house. If you look at the house as a piece of sculpture, the interior furniture is meant to complement, not distract from it. Then there's the secondary sculpture—the fireplaces, the kitchen—and our pieces of Inuit sculpture. It's really a "patron of the arts" sort of thing.

We started in January of 1999, and it was done in the fall of 2006. Almost seven years. But I think what is really amazing there was no conflict among the artists and the patrons. Everything worked smoothly. You get inventors, sculptors, painters, all collaborating on one project, and it's definitely offbeat. When you see the art, it crystallizes for you.

We wanted unfolding views of the land, certain materials like slate, with muted colors. We wanted an indoor-outdoor feeling, rooms that could transform into bedrooms, next to bathrooms, not half baths, and a yoga room with a bed that could come out of the floor, for a bedroom. We wanted this whole combined living room/dining room so we could have fundraisers. We've had 120 people over for a wedding.

The program for the house would be on four floors, with a gym at the bottom and the attic, to be Eileen's offices. My office would be on the ground floor, along with the living and dining areas, the kitchen, and a porch. We had the same amount of rooms in Belmont, twelve, but here we didn't want to see the garage. That would be in the basement. We wanted to make sure the landscaping did not have a blade of grass. Moss, ferns, indigenous flowers, and trees. The fencing would be a piece of sculpture. Everything was thought out in terms of artistic orientation. It still had to have a facility to complement our lifestyles. We both work at home ninety percent of the time.

We met every week with the architects at the beginning. At first they would propose something, and we'd reject it, or tweak it. At one point, frustrated, we told them to think outside the box: Why not sculpture? They were making red-brick fireplaces, and we said, "Just break out of the box." They said it's going to cost a lot of money, and we said "fine." So they got it and broke out. We revised some of the original stuff, like the fireplaces. But here's the other thing. Of all the materials in this, there's no gold and no marble. It's slate, it's concrete, it's cast-iron, it's aluminum foil, it's bluestone, and steel and glass. It's all just basic materials. There's nothing fancy in this house. The staircases were made out of stainless steel. Each piece harmonizes.

We wanted an interior designer that wasn't an interior designer by trade. We wanted an architect who does interior design. Calvin Tsao was recommended by Diller Scofidio, the architects of the Institute of Contemporary Art. And we said to Calvin, "We want you to complement this layer with an inventive interior design." We'd go down there to New York and he'd give us models, and we went around looking at stuff. I rejected only one thing: the dining-room chairs they designed, because to me they were too clunky. I went to a store in New York, sat in one, and liked it. It's very exotic and sensuous.

In the attitude of the house to the site, we wanted it to be dramatic, to hang over the precipice, over the pond, but in the trees, like a tree house, so views of the pond would be accessible from every room. We also wanted it sited so the sun coming in hits at different angles during the day. That's why there's glass front and back. We told the landscape architect that the most important thing was that the landscaping not look manmade. Just plant the kind of things that are in the woods. If you have birch trees, put in more birch; moss, put in more moss. As far as the fencing goes, make it look like outdoor sculpture. We didn't want a manicured landscape. The flowers, not in rows. We wanted them to look helter-skelter. The first question I asked our landscape architect was who was her favorite composer, because she had been a musician, a classical pianist. It wasn't Bach, but I don't recall who it was, because I wasn't so interested in what she liked, but why she liked it. I wanted to hear her articulate it. I don't know why she gave up the piano. I think she told me and I forgot. I thought she'd be appropriate, because she'd bring a more artistic interpretation to the landscape. And she was a sculptor and had a direct experience with the arts, not just somebody who went to design school.

Music. I think it's, by nature, collaborative. You play with other people; and as a musician, I have that instinct with me. There's always this sensitivity to what they're doing. I told our son, "You've seen the house—write something that you think appropriate." It's interaction. I didn't tell the sculptor what to do. We didn't want the landscaping to be any different from what the furniture people, or the architects, or sculptors were doing. Don't take three acres and put grass. It doesn't fit in. Make it harmonize with this beautiful landscape. There are ten acres of land, five contiguous and five across the pond. At the very least, we're thinking of growing grapes there. We have no intention of developing it. We were given an envelope, a building envelope. We built within it, and there was more than enough space. It's 9,500 square feet, and with the garage and gym in the basement, it's 14,000.

The whole thing was a process of discovery for us. You kind of knew there's a form of architectural expression, and that its manifestation would not be static. Instead of looking at it, it's the experience of walking through it. The other day, I was seeing the sky from the atrium, but it was the reflections of the sky on the ceiling kitchen. I stepped into the kitchen, and it was like stepping into the mirage. The aluminum leaf on the whole belly of the ceiling reflects all the different light, and dances off the ceiling. The view is unfolding all the time. It's the most stunning house when you're there and the sun is going down.

Since we don't practice the visual arts—we do that socially—and it made a lot of sense to do the house this way since they're our friends. We've got a house that's represented by twelve artists who have all marked the house in their own way, all trying to complement, not compete. As we get older, I think the house will help me deal with the human condition. It has a very sedative composure to it; there's a very calming feeling. It's like the way people lived a thousand years ago, living where the light is—where I want to sit with a guitar and read a book, where we plan a meal. And it'll be a more creative environment to do music. The acoustics are really good. It resonates.

Here, of course, so close to Walden Pond, you think about Thoreau and his existence in the woods. Not in the same poetic way as Thoreau. But living in the woods is calming. Certainly, the house and land have a very spiritual feel to it, almost like a sanctuary. I'm very spiritual, but not "deity-based" spiritual. I'm more "nature" spiritual. God, to me, is nature. So when I die, I'll sink down and nestle into the pond. And you're really communing with nature as you approach older age. I couldn't think of a better way to approach nature, with flora and fauna and the pond. It'll help me deal with mortality.

You can't see another house. The pond is two miles long. You can only swim and take canoes out. Lincoln has no restaurant, and a quarter of the land is in conservation. It's a very rural town. It's verboten to have boats on the pond. We love the city, and we love the country, and we hate the suburbs. And unfortunately, until now, we've spent all of our lives in the suburbs. When we brought this house to the Planning Board, there was a brouhaha, and we had thought Lincoln was a haven for contemporary architecture. One of the guys at planning asked, "What color is it really going to be?" and we said we wanted to blend into the landscape. And the house behind us, it's traditional, but big and white, sticking out like a sore thumb.

Walter Gropius, who built his famous Modernist house here in 1938, is pretty much forgotten. The reason is that most people who come into Lincoln build one of these awful McMansions. So without looking, the planning department said big must be bad. It took us a year to get through the Planning Board. The house, they said, was too close to the edge of the water. We tried to be accommodating. We ended up moving it back ten feet, but refused to move it back twenty. "Too big," they said. We said, "Hey man, you gave us the building envelope." And they said, "You filled up more than we expected." If they got cute with us, we would have sued.

Minutes, Planning Board,
Wednesday, December 15, 1999, Donaldson Room
Preliminary meeting for site Plan Review

9:45. The Davolis own 3 lots, want to build on one lot, within the building envelope. Total above grade 9470, 3 stories above grade, height 35 ft. Modern design, landscaping to minimize effect of lighting. Gordon Winchell and neighbors are present. Concern about cutting trees on the Pond side, views from the trail. Plan for site visit, Jan. 8, 8:00 a.m. 2 issues from Gordon – trail from Huntley Lane to pick up existing trail and water easement loop. Also they want to move the trail on the first lot. We need to look at that.

Planning Board Minutes, January 19, 2000

Owner Robert Davoli appeared along with architect Warren Schwartz. Mr. Schwartz presented an update of the plan as well as a model of the home. He stated that this was a personal vision for the homeowner and outlined their compliance with the intent of the bylaw:

- Preservation of natural forest and ground cover,
- Removal of trees only for the house and septic system,
- Minimum changes to the topography,
- House massing and roof lines which follow the natural contour,
- Use of colors and texture of exterior materials which blend into the site,
- No areas of lawn created,
- Open area of septic field will be planted as a wild meadow,
- Garage will be concealed underground,
- Discreet site lighting.

Discussion turned to lighting. Mr. Schwartz commented that the pond side of the house is always in shadow. They intend to plant extensive screening with indigenous trees. The narrow elevation of the house faces the nearest neighbors. The house was sited within a building envelope developed by the cluster approval process. The house takes up 10.9% ratio of the 2.37 acres. Comments on the lifestyle of the owners included a commitment to the arts and the environment. Steve Stimson, [then] landscape architect, presented a plan indicating existing trees and proposed additional landscaping to address concerns of the neighbors. Some 46 trees will be removed from the building site; 92 trees and 56 shrubs varying from 8' to 28' in height are proposed to be planted as specified. Mr. Schwartz presented photos taken on the day of the site visit, with computer renderings of the house superimposed onto the site. A more traditional home was also superimposed on the photographs. Neighbors Peter and Ellen McCann said they have concerns with the relative scope and perspective of the proposed house. They also said they are encouraged tonight by the presentation of plans which are specific regarding landscaping and which preserve views to the conservation land. The McCanns have retained their own landscape architect to help them understand these plans. They hope that the final landscape plan will be incorporated into the Planning Board approval. They also have a concern as to the scope and scale of houses which could be built on the other adjacent open lots, and with the amount of glass in the proposed house.

Buzz Constable, Lincoln Land Conservation Trust, spoke to the interests of the LLCT and trail users, and wondered if the house could be pushed back 15 to 20 feet. He hoped that the Planning Board would enumerate the conditions of the approval such as the colors, lighting, etc. The Breslins said they want to maintain trails as they presently exist. The Board resumed their discussion. Mr. Cooper asked to review calculations of the mean grades around the house and elevations. Mr. White praised the presentation, but wondered if the Board was being lulled into a false sense of security regarding the project. Mr. Schwartz said the form and materials are intended to break down the size and scale of the house. Mrs. Faran said she is more concerned with the views from the trail. Mr. DeNormandie said that the bylaw speaks to not seeing the house on the most prominent site. This closeness to the edge of the slope needs to be mitigated. He also noted that at night the windows will spill out light. Screening is one mitigating factor; but there needs to be other ways to cope with the expanse of windows. Mr. Constable asked if the Board is receptive to adjusting the building envelope. The Board noted it is within their jurisdiction. Noting the very large size of the house, Mr. White said he would like to work with the applicant regarding the entire site (three lots). Gordon Winchell said he is comfortable with the location of the house, noting that it complies with the original cluster, that it is not that different from the condominiums located on the pond, and that he feels more reassured having seen the presentation. Mr. MacLean asked the architect to scale a person on the elevations to get a sense of the height. A motion was made and seconded to continue the hearing to February 16, 2000 at 8:00 P.M. The Motion passed unanimously.

concept

1 concept sketch

warren schwartz **Principal, Schwartz/Silver Architects**

I met Bob and Eileen in 1997. They called looking for someone to design a house, and wanted to come over to the office and meet me. My first impression was that Bob was dynamic, and Eileen was bookish; I thought that, as a couple, they were interesting, and I wondered how it works. They went away for about a year and came back, having put together a piece of property. They had visited other architects, and they reported to me that the reason they hired me was that ultimately Bob sees himself as an aspiring musician, but feels like a failure. He'll agree that he knows how to make money. "Really," he says, "I'm a failure because what I really want to be is a musician."

This house is something that could not have been done without the two of them. Eileen says, "I'm a writer, and Bob's a venture capitalist." But I said, "You're both artists in different skins." Anyone who wasn't an artist couldn't produce this house. Developers are very good at putting people in pigeonholes, but not Bob. He's one of those people who wouldn't ever call a "starchitect." The reason they would hire me was because I could live with a musician (my wife) for all these years—Eileen's a classical music lover, and he's folk and blues, and if you add their sons, one's into Bach, and the other, Robert, writes music. And they all perform, except for Eileen.

Bob met Eileen when she was a recently divorced mother of two sons, and he was a short order cook at Milton Academy, and took a job as a baby-sitter for Eileen. Bob eventually grew to become their emotional father. As far as they're concerned, he's their dad. I find such information so interesting when I begin a project. I want to know who my clients really are, what motivates them, how they see their own lives and how they see their lives unfolding in relation to others' lives, how they see themselves in society, and how they see themselves personally. I think it's absolutely critical to know a little too much: to know who they are together, and who they are separately. I had to feel as though I know them, because a house is the most personal of projects to design.

From Bob's point of view, it wasn't about style. He wanted a house woven into the woods. He has paintings, and sculpture—mostly Zimbabwean and Inuit. Basically, carved stone. He showed me some of the work. Both collections of sculpture were dark, if not black, in color, and this had something to do with the eventual color and form of the house, because it was, in a sense, a handcrafted work, and it was to house these other works. They had art objects with a social and cultural background, rather than, say, industrial. Bob and Eileen were particularly attracted to handmade aesthetics. They really prize the human touch, and I thought the house itself should be a little like what they like in the art they collected, but different. The house they wanted was large and complex, but not necessarily high precision; the challenge was to make this a very precise work (because it would have to be very precise in construction), but not to let the precision of the work preclude the hand touch. That was important. They also seemed to like natural materials, unpainted things. And so I began thinking about walking around the woods, and it would be like finding an object in the woods.

When they called back, I got into a car with Bob, and he took me to the site. It was incredible. One could imagine putting the house in the woods, not just on the pond. Basically, it would be a found object—a part of the woods, not foreign to it.

So on the way back from seeing the property, he said, "We'd like you to do the house." That was the moment. Bob is a very decisive person, and so is Eileen. I was actually a little surprised, and I felt challenged by that, but I also found it was a fantastic opportunity to do something extraordinary. At that moment, I hadn't felt as though I'd been competing for the project. It was just something that I'd wanted to do. I didn't know that there were people who are so unusual, that they'd want me to do something like this. Few people want a work of art. But no one knew where it was going to go. They didn't have many rules, but when he said "not a blade of grass," he said that it meant he didn't want a "McMansion," a trophy on a green-velvet pad. The point was not to be large or French provincial or English country, nor to come from the outside in. For me, this was, in a way, perfect. Bob and Eileen were inviting me into their psyche. That was so much more important, so much more collaborative, somehow so much closer to the core of why you become an architect.

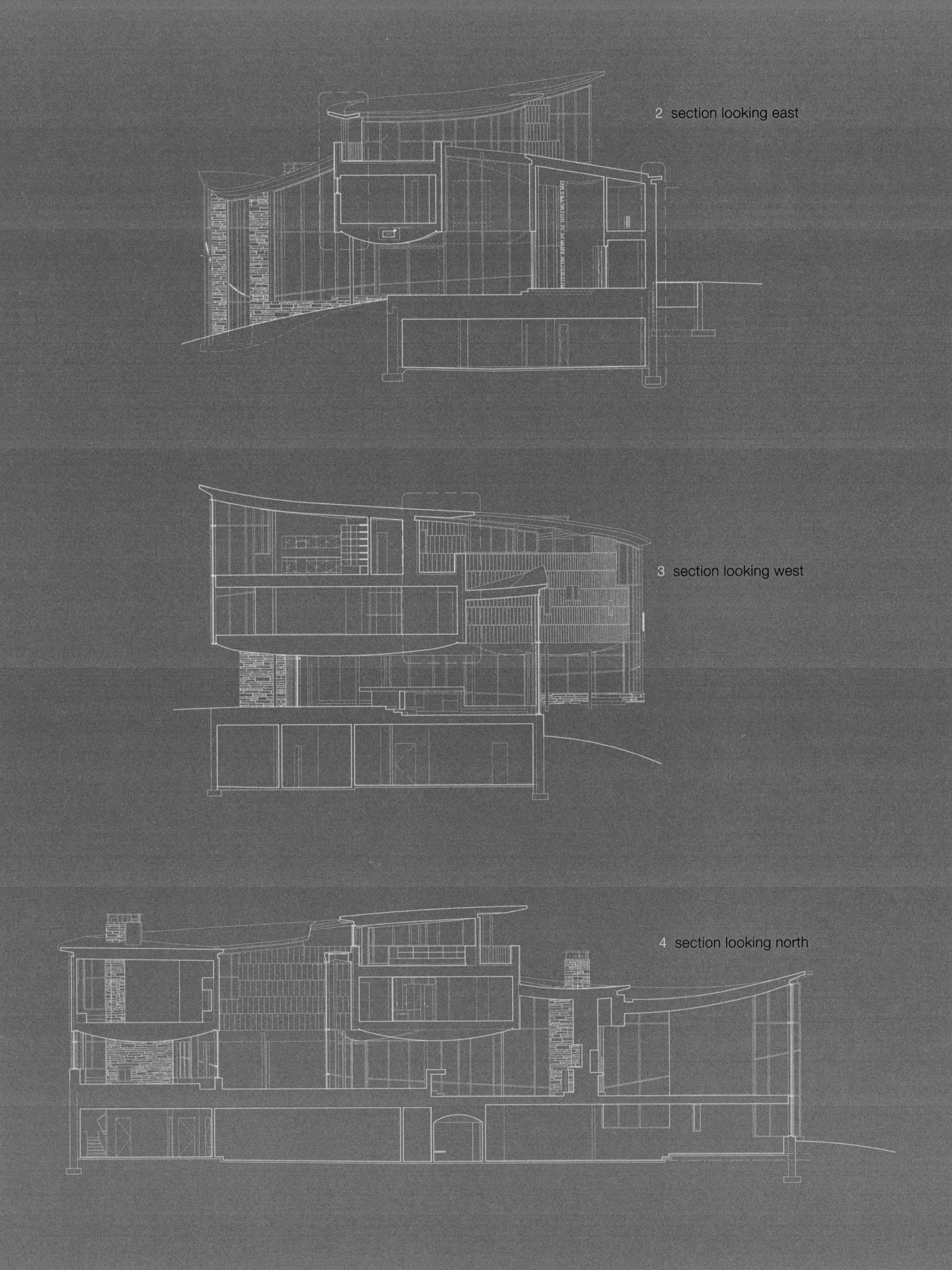
2 section looking east

3 section looking west

4 section looking north

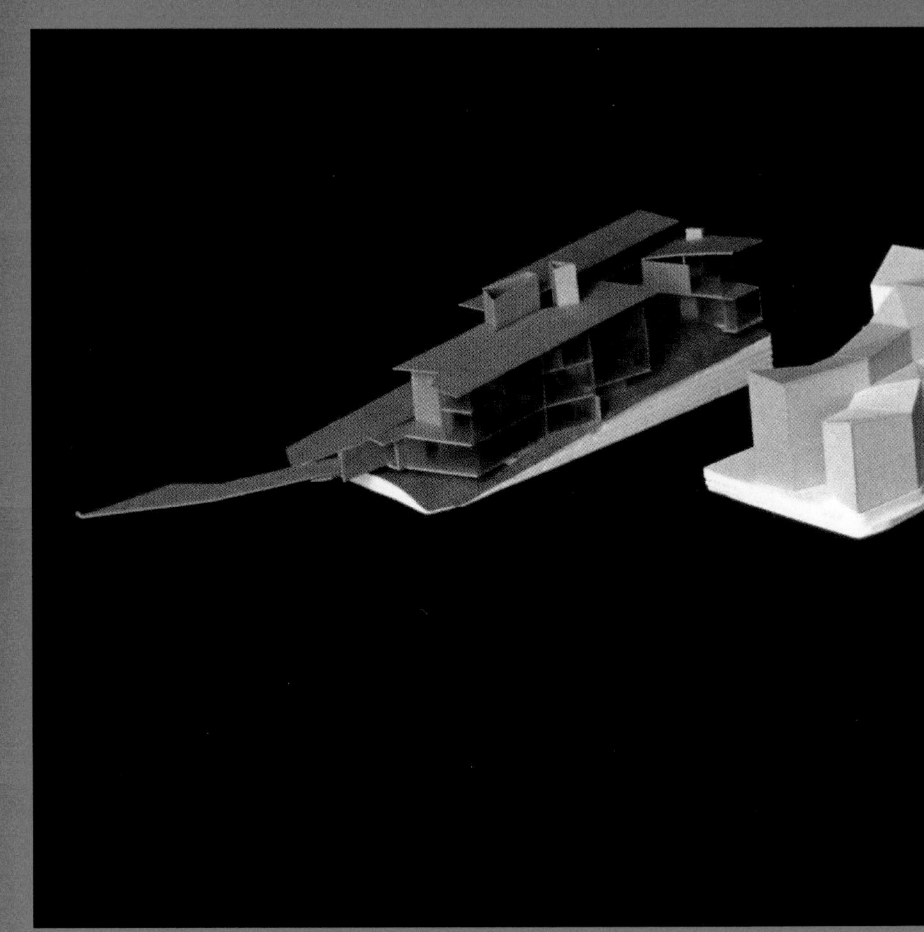

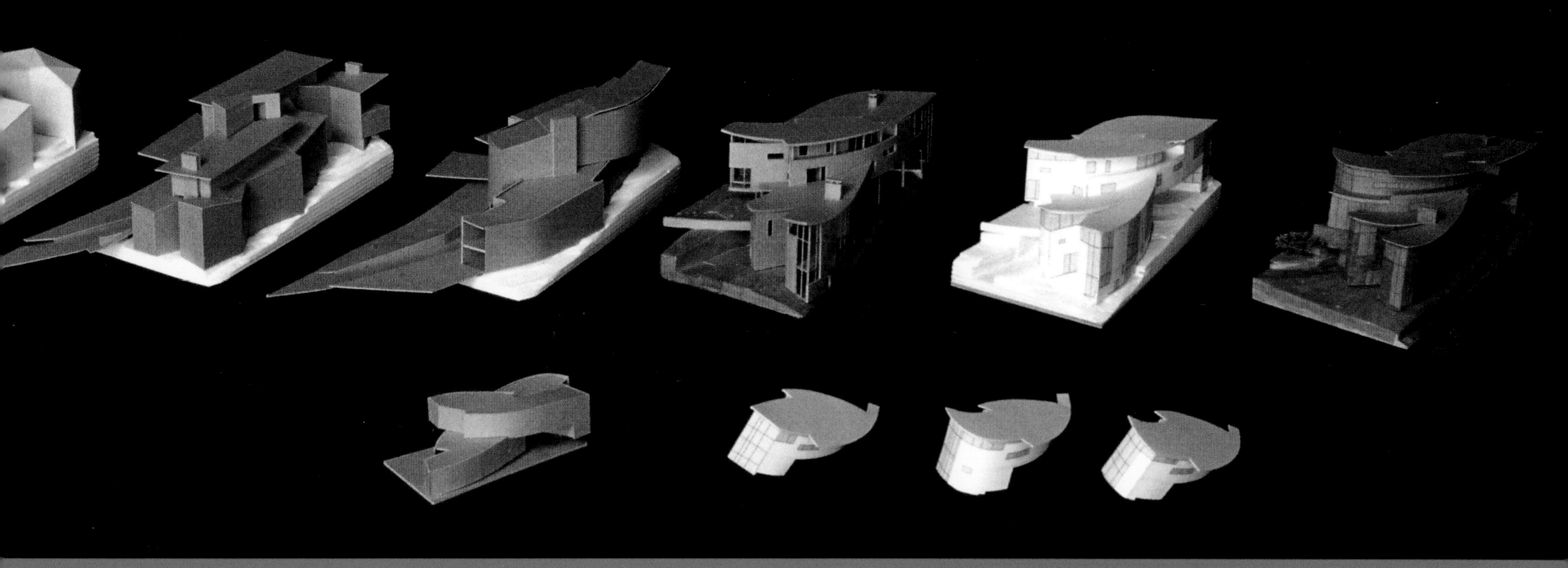

5 the design developed through seven sequential models shown left to right. the small chipboard model marks the turning point in the design, represented by the concept sketch on page 23. the small white models show development of the upper floors.

6 computer models of the final design

gordon winchell Conservationist and Neighbor

We moved here in 1925, when I was six, probably in grammar school. The Farrars had owned the land since the 1600s, and EdMr. Farrar, who was considered a ne'er-do-well by the farmers here, created Farrar Pond. He was in his sixties. A stream called Pool Brook went through the land, curled up, and went into Fair Haven Bay. They used to hay the land but when the river was high, the whole area got flooded. All you could do was look at the water when it was high. The only thing Ed Farrar had to do to make it into a real pond was put in a 300-foot-long dam levee and clear the trees. That was a hundred years ago.

The lake is now in conservation land, managed by a conservation trust for the benefit of the town. The canoes and rowboats are owned by abutters of the pond, and supposedly the fishing is for the abutters. There's no easy way for others to get onto the pond. Every three years, we draw down the pond four feet, which puts most of it out of water. The deepest anywhere is about eight feet, but it's mostly four or five feet, with a lot of mud. Our main interest is to keep out the weeds, and keep it from becoming swamp and brush. We draw it down and let it freeze and let water come in, which reverses the production of lily pads and gets rids of nutrients, which flow in. We try to keep out invasive plants. So far we've been able to maintain the pond, which is in as good a shape as it was in 1940. We will be doing one this spring, because of Eurasian milfoil. This is one of the major things our homeowners deal with: trying to keep their property in good shape. We look forward to having the Davolis participate.

When I was a boy, I went around with Mr. Farrar, who showed me a hatchet mark that was put in by Thoreau when he was surveying. So there is this protective feeling about the place, and a sensitivity to building here. Many of the old families that have the land want to preserve it as much as possible, to try to keep land use in the character of the town. That's why there's so much open space.

Our family bought the land from Mr. Farrar when he wanted to sell the south end. They bought 200-plus acres, and we've been trying to follow his goal of keeping a natural lake. When we decided we had to do something with the land, we favored a new type of open-space zoning, which allows clustered zoning rather than two- acre home sites. Two sets of condos were built for people who want to stay in town and sell big houses. So we're happy at having to keep as much open land open in conservation, including the pond.

We didn't know what the Davolis were going to build, what they had in mind when they bought the property. I don't have any clear memories of the discussions at the hearings, though I went to some. We had made a permanent trail crossing their site coming down to the lake, and Eileen wanted to eliminate the trail, so it wouldn't cross the driveway. She didn't like the trail on the south side of the lot. But she agreed to have it cross the driveway, and the house looks reasonably good from that angle. To have the trail cross the driveway, the state legislature had to vote on their approval.

When we set the property up for sale, we established the building envelope, and they pretty much filled that up. It's, of course, a very large house, and in Lincoln, we're worried about McMansionization. I was especially worried about how their house would look from the pond, but there were restrictions in place. If people own their view, and clear cut down to the lake, then everyone looks up and sees a big white house. We've been trying to preserve the shore of the lake, to save the frontage. Most of the houses around the pond are protected by the pond, so you don't see big-faced houses. A couple of houses are disturbing, but most respect the pond. When you're out on the pond, the Davoli house doesn't stand out. The tall trees are thinned out a bit, with white birch dying. It's okay if they replace the birch, as long as it's native birch, not the European variety.

People were struck by the size of the house, but not too discomfited. The shape of the house is very compatible with the shape of the rounded hillsides. It's not like new houses, with their white peaks. The stone exterior fits in well, and so does the color. So from the standpoint of the surroundings, it works remarkably well. The fence they have around the house is unique. Trying to keep the deer out, or the llamas in. Originally, Eileen was trying to raise llamas. It's an artistic thing that goes along with the brave architecture. Considering the issues around building the house, I'm comfortable with what they've done, and they respected the contours of the project, from the siting to the sight lines. Then they bought the adjacent lot, and another, so now they have three lots. What's they've done fits.

7 lower level

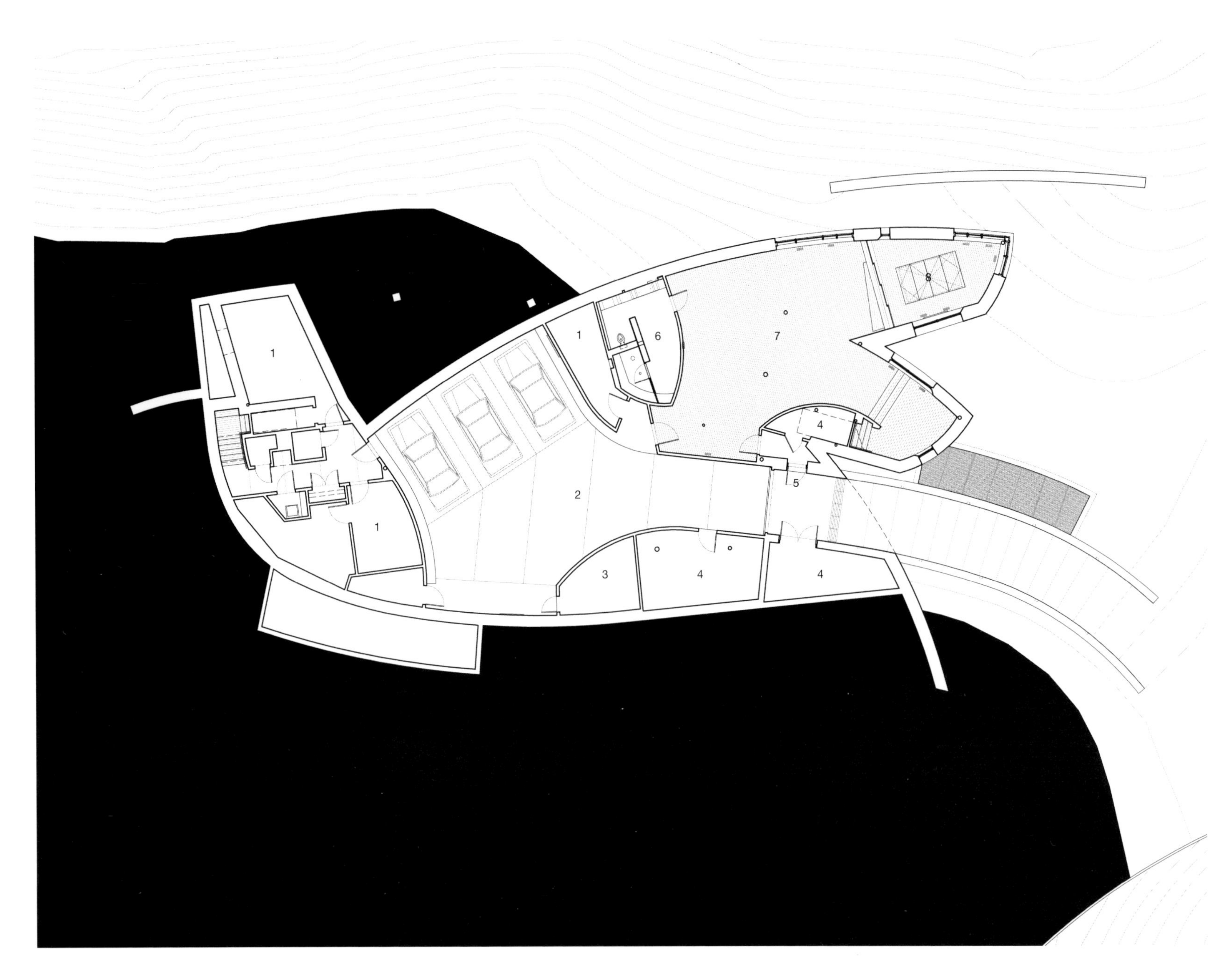

1 mechanical
2 garage
3 wine storage
4 storage
5 secure entry
6 bath
7 exercise
8 yoga

8 main entry level

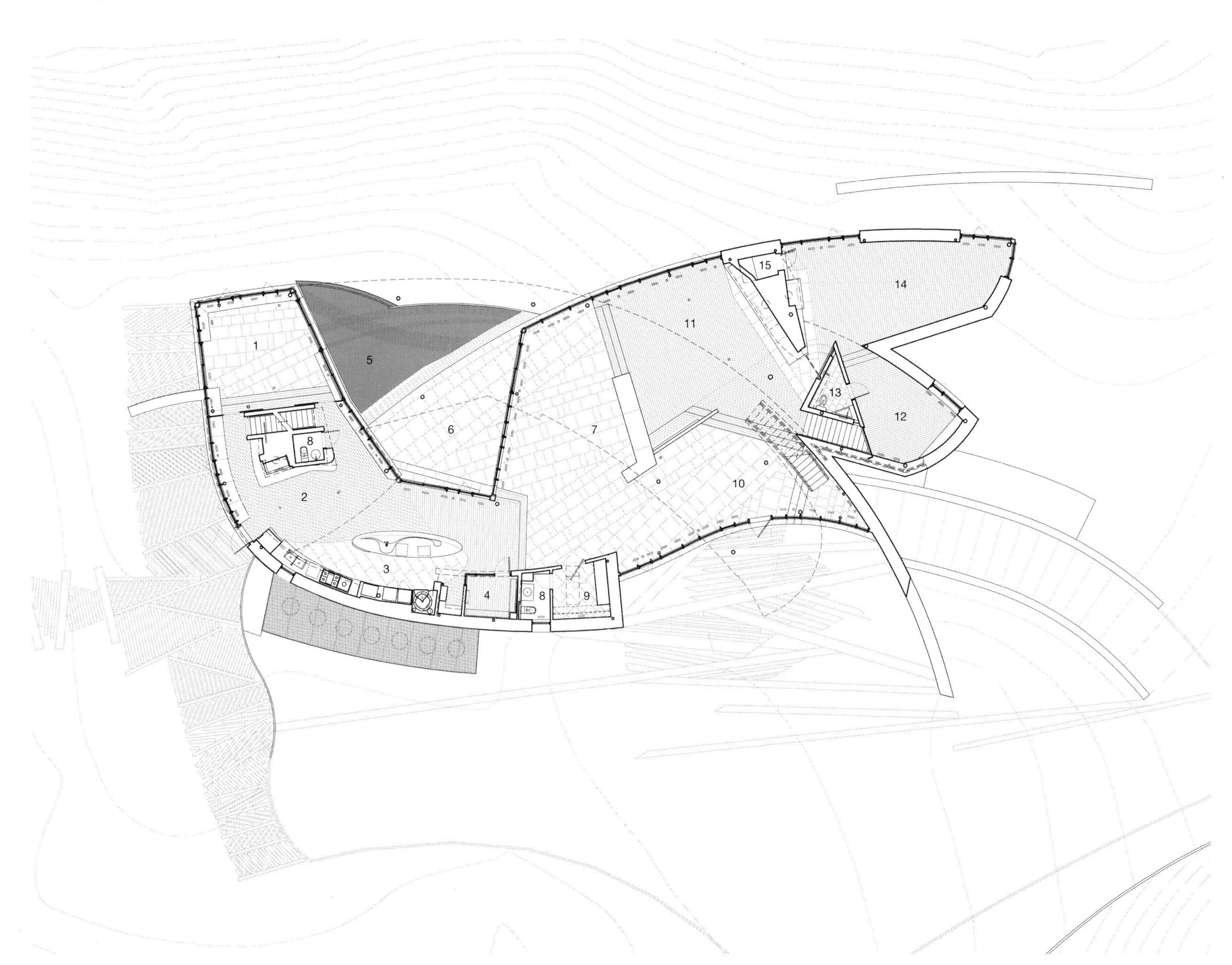

1 porch
2 sitting
3 kitchen
4 wine
5 koi pond
6 court
7 dining
8 bath
9 entry closet
10 entry
11 living
12 conference
13 bath
14 office
15 closet

9 second level

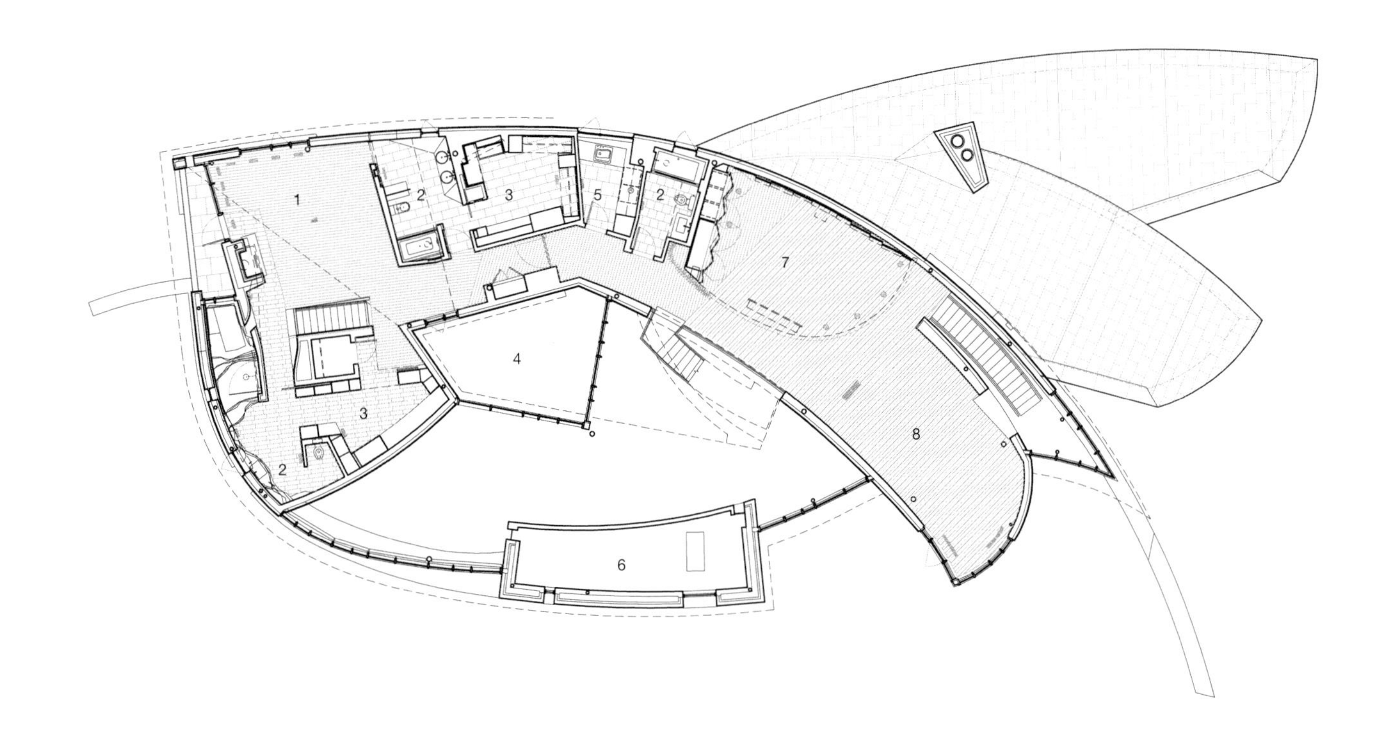

1 master bedroom
2 bath
3 closet
4 void
5 laundrey
6 mechanical
7 media
8 family

10 third level

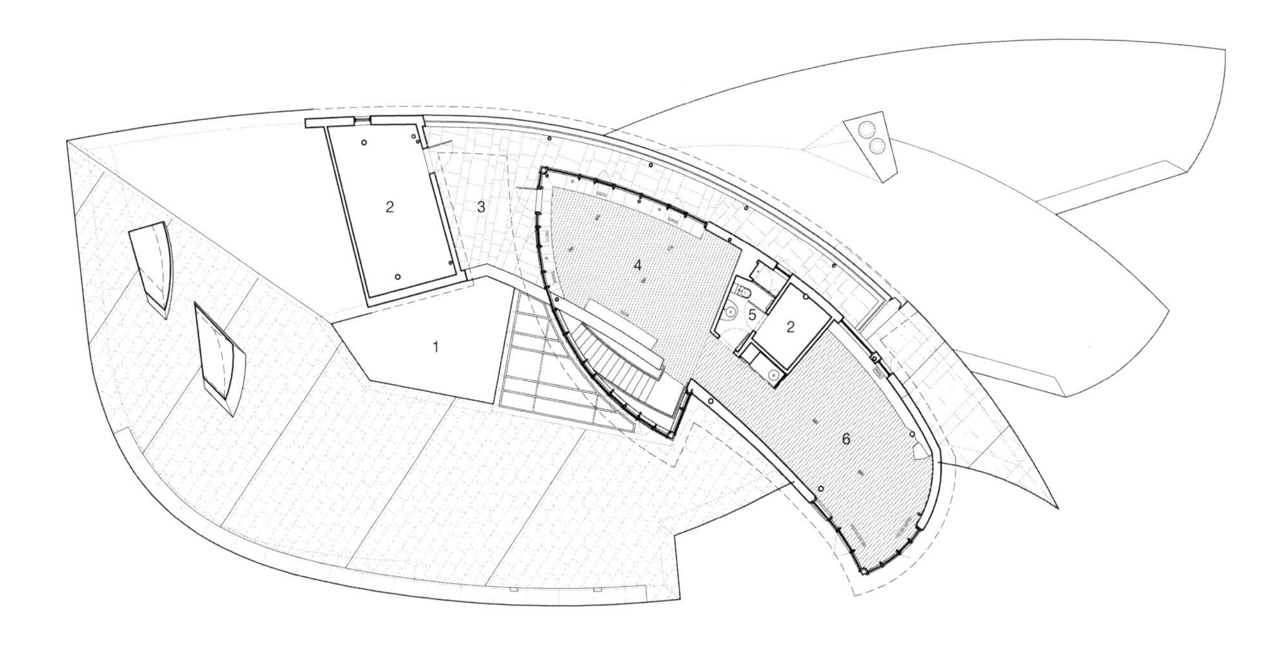

1 void
2 mechanical
3 terrace
4 dayroom
5 bath
6 office

11 final site plan

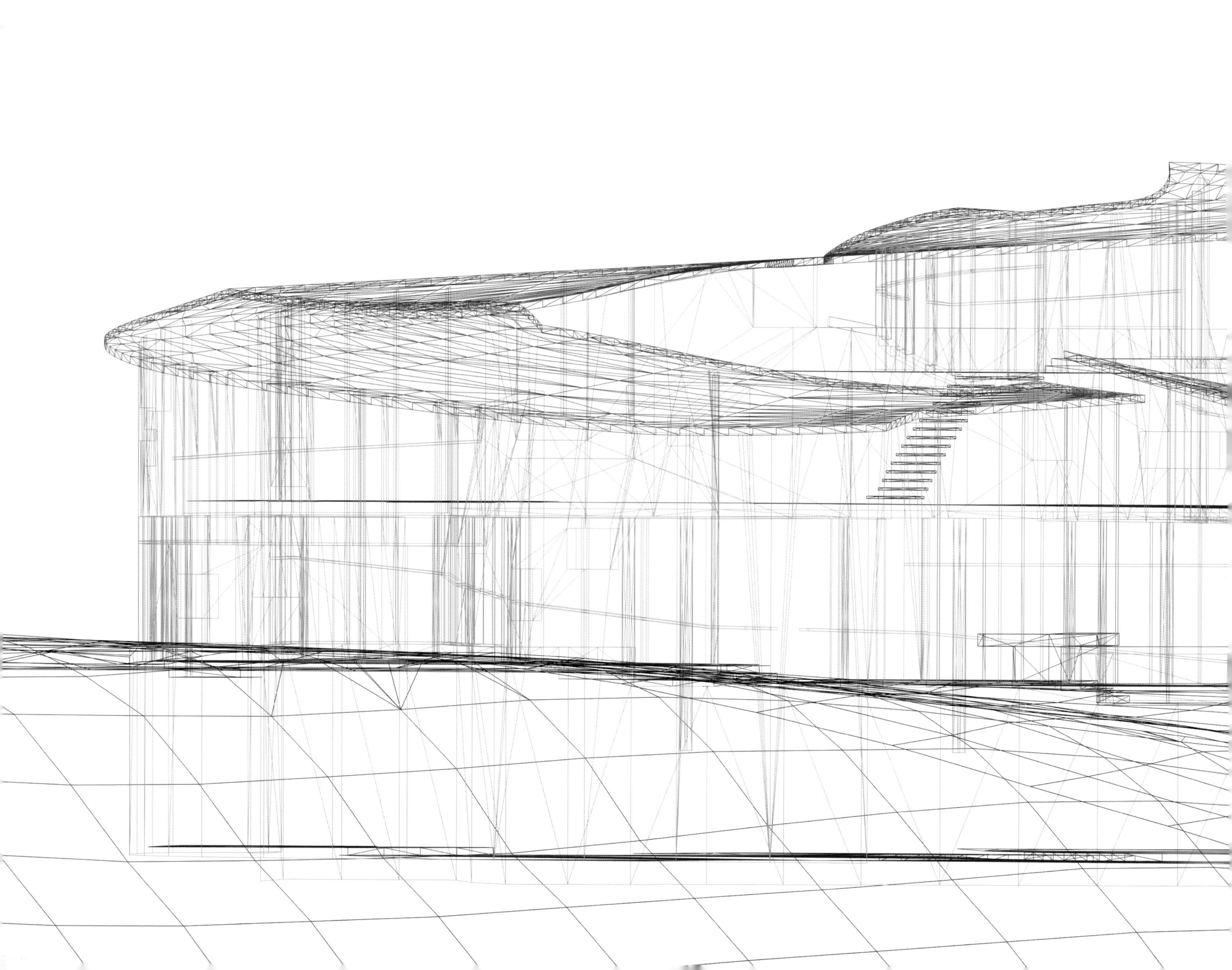

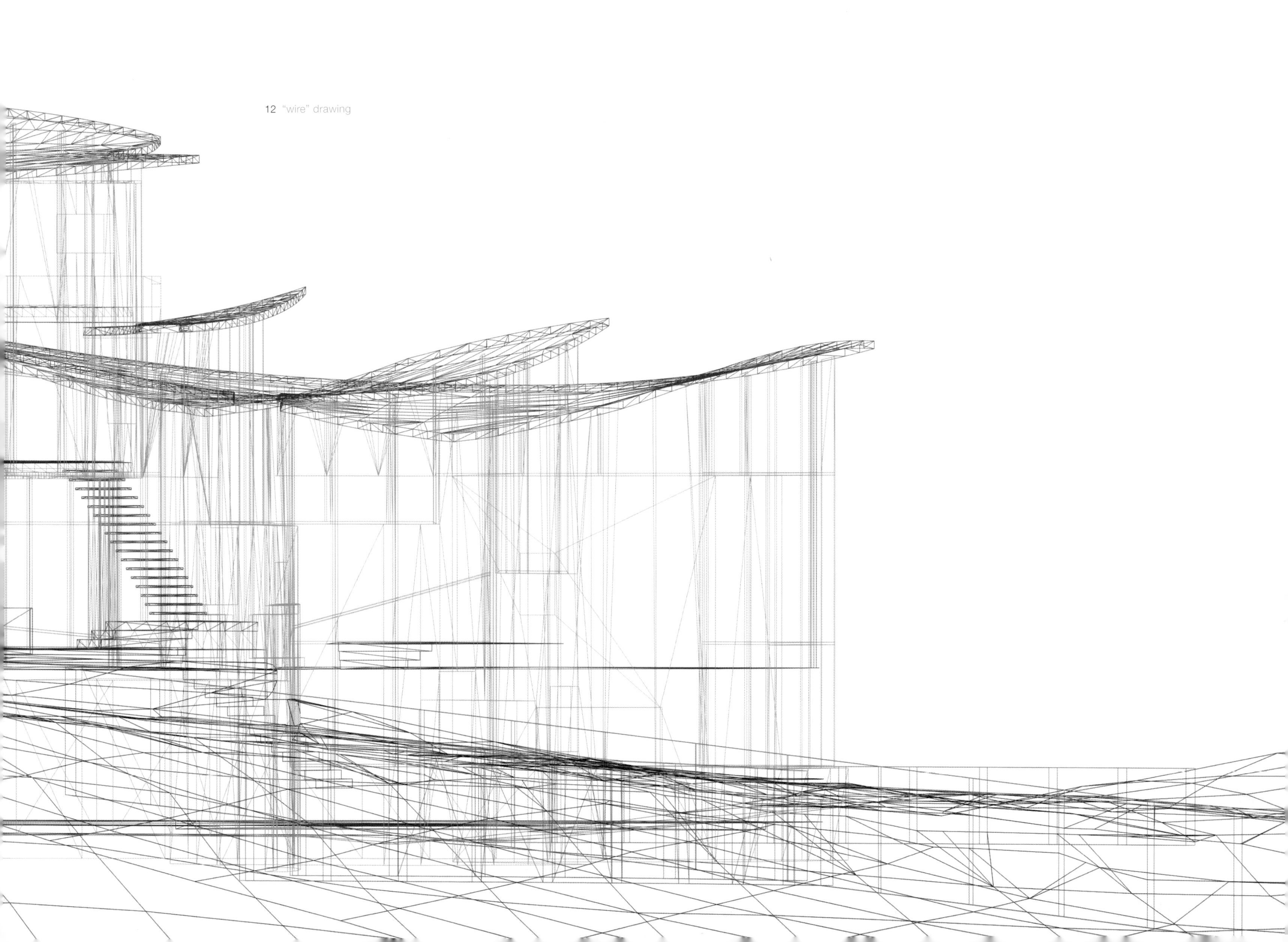

12 "wire" drawing

13 farrar pond

henry david thoreau *Walden*

A lake is the landscape's most beautiful and expressive feature. It is earth's eye, looking into which the beholder measures the depth of his own nature. The fluviatile trees next the shore are the slender eyelashes which fringe it, and the wooded hills and cliffs around are its overhanging brows.

Standing on the smooth sandy beach at the east end of the pond, in a calm September afternoon, when a slight haze makes the opposite shore-line indistinct, I have seen whence came the expression, "the glassy surface of a lake." When you invert your head, it looks like a thread of finest gossamer stretched across the valley, and gleaming against the distant pinewoods, separating one stratum of the atmosphere from another. You would think that you could walk dry under it to the opposite hills, and that the swallows which skim over might perch on it. Indeed, they sometimes dive below the line, as it were by mistake, and are undeceived. As you look over the pond westward you are obliged to employ both your hands to defend your eyes against the reflected as well as the true sun, for they are equally bright; and if, between the two, you survey its surface critically, it is literally as smooth as glass, except where the skater insects, at equal intervals scattered over its whole extent, by their motions in the sun produce the finest imaginable sparkle on it, or perchance, a duck plumes itself, or as I have said, a swallow skims so low as to touch it. It may be that in the distance a fish describes an arc of three or four feet in the air, and there is one bright flash where it emerges, and another where it strikes the water; sometimes the whole silvery arc is revealed; or here and there, perhaps, is a thistle-down floating on its surface, which the fishes dart at and so dimple it again. It is like molten glass cooled but not congealed, and the few motes in it are pure and beautiful, like the imperfections in glass. You may often detect a yet smoother and darker water, separated from the rest as if by an invisible cobweb, boom of the water nymphs, resting on it. From a hill-top you can see a fish leap in almost any part; for not a pickerel or shiner picks an insect from this smooth surface but it manifestly disturbs the equilibrium of the whole lake. It is wonderful with what elaborateness this simple fact is advertised,—this piscine murder will out,—and from my distant perch I distinguish the circling undulations when they are half-a-dozen rods in diameter. You can even detect a water-bug (Gyrinus) ceaselessly progressing over the smooth surface a quarter of a mile off; for they furrow the water slightly, making a conspicuous ripple bounded by two diverging lines, but the skaters glide over it without rippling it perceptibly. When the surface is considerably agitated there are no skaters nor water-bugs on it, but apparently, in calm days, they leave their havens and adventurously glide forth from the shore by short impulses till they completely cover it. It is a soothing employment, on one of those fine days in the fall, when all the warmth of the sun is fully appreciated, to sit on a stump on such a height as this, overlooking the pond, and study the dimpling circles which are incessantly inscribed on its otherwise invisible surface amid the reflected skies and trees. Over this great expanse there is no disturbance but it is thus at once gently smoothed away and assuaged, as, when a vase of water is jarred, the trembling circles seek the shore, and all is smooth again. Not a fish can leap or an insect fall on the pond but it is thus reported in circling dimples, in lines of beauty, as it were the constant welling-up of its fountain, the gentle pulsing of its life, the heaving of its breast. The thrills of joy and thrills of pain are undistinguishable. How peaceful the phenomena of the lake! Again the works of man shine as in the spring—ay, every leaf, and twig, and stone, and cobweb sparkles now at mid-afternoon, as when covered with dew in a spring morning. Every motion of an oar and an insect produces a flash of light; and if an oar falls, how sweet the echo!

In such a day, in September or October, Walden is a perfect forest mirror, set round with stones as precious to my eye as if fewer or rarer. Nothing so fair, so pure, and at the same time so large, as a lake, perchance, lies on the surface of the earth. Skywater. It needs no fence. Nations come and go without defiling it. It is a mirror which no stone can crack, whose quicksilver will never wear off, whose gilding Nature continually repairs; no storms, no dust, can dim its surface ever fresh;—a mirror in which all impurity presented to it sinks, swept and dusted by the sun's hazy brush—this the light dust-cloth—which retains no breath that is breathed on it, but sends its own to float as clouds high above its surface, and be reflected on its bosom still.

A field of water betrays the spirit that is in the air. It is continually receiving new life and motion from above. It is intermediate in its nature between land and sky. On land only the grass and trees wave, but the water itself is rippled by the wind. I see where the breeze dashes across it by the streaks or flakes of light. It is remarkable that we can look down on its surface of air at length, and mark where a still subtler spirit sweeps over it.

Warren Schwartz, *continued from page 21*

Eileen said they didn't need the house, that it was an art project, but with a purpose. If they support all these other artists, they could go the ultimate step and immerse themselves in a work that surrounds them. Not just seeing it from the outside, as an audience, but really becoming part of it.

One of the first ideas of the house was something about a courtyard, though no one called it that. Somehow we had to bring the outside in, so that literally, when it's raining or snowing outside, it would be raining or snowing inside. You should feel the sun inside. The idea of a pavilion in the woods was also emerging in the discussion.

The town of Lincoln prides itself on its rural nature, and there are many walking paths through everyone's property. You can walk on the path, which is a kind of easement, so you're constantly finding people walking through other people's lands. We knew the house would be in the public eye. Of course, the Gropius House was in the back of my mind, too, throbbing. It had recently appeared on a stamp. Both Eileen and I were aware of the Gropius House, and the values it brought to the area fifty years earlier. It didn't have much to do with the siting of house, or the colors, but the aesthetic role. How would something like that be approached now?

That was part of the challenge. You're always living with context, and the context is always selected by what you choose to relate to. In this case, the choice was to relate to these early Modernist houses, which ultimately made their way into a historic setting. I really did want to put it in that context, and everyone did. It was more than a house for Bob and Eileen. They felt that way, too. You could refer to it as a house, but the art of architecture was expanded to include as many arts as possible, performing as well as fine arts—their intention was to have concerts there. She thinks it'll have the power to change the way people spend their money, into endeavors more worthwhile than impressing friends.

At the beginning of design, they asked for a lot of things. They wanted to put their entire life into the house, all the things that kind of defined who they were and what they wanted to do. So the house got a very diverse and extensive program. I interviewed them first to find out not how many showers they had, but what they thought about what they did, to get a sense of what their hours and days and weeks were like, to get a sense of what the house wanted to be, and the way they'd be living into the future. It included things like Bob's wine cellar, where he'd have something like seven different kinds of wine glasses to serve from. I think of him as a first- or second-degree black belt in wine, as thought he's got a doctorate in how it grows and smells. He wanted a place for his two dozen guitars, which he collects and plays. He needed his office, which has the guitars in it, and a conference room, for people who come over to have their meetings. Eileen needed an office for herself, and a place for tons of books, a place where she could almost live with her books. She spends a tremendous amount of time writing. I viewed Eileen's perch as a crow's or eagle's nest, something way up high, in the trees, so that she can be highly focused when she works. Bob, who is tremendously gregarious, has a lot of people he communicates with, and Bob wanted his office to be close to the front door.

"What about laundry?" I'd ask. They were capable of talking about it, but not interested in it. Instead, they were interested in the aesthetic quality of the house, of the spaces around their activities. That's where they stopped being prescriptive. Their instruction to the jury was to do your best work, aside from program and size and location, and to make it a challenge to be creative about. They liked to entertain many different ideas, and I kind of view it as a house filled with many ideas, even converging and diverging ideas. Because it's what their lives are like, and it's what most people's lives are like—successes and failures and averages.

The house challenged our own abilities to work with others and find a form which would relate to their interests and our own. In other words, the house didn't spring only from the idea of the architect. In fact it was an amalgam of ideas, coming from the clients, the site, and our own position. They wanted to bring all of this out, and they both liked natural crafts, but that extends to pretty sophisticated crafts, like a ring collection, which are actually sculptures on a miniature scale. Some are modern, some are representational, some are stories. I don't know where she gets them. She doesn't have many collections, just rings and books; and he, wine and guitars. They choose the art together. They come to a consensus about the art they buy. In their collection of stone sculptures, of Zimbabwean and Inuit art, they like the notion of movement in the black carved stones.

Bob is always away during the week, but I don't know of a time he wasn't home over the weekend, and they'd spend the weekend together, and they'd get rid of the work week by relaxing, and the two would do social things. They asked to have a screening room, which figured prominently in the program, and so did the billiards room. They wanted a place where their sons, Robert and Edward Price, would come. They both exercise, so there's an exercise room below.

When you have a billiard room, and offices, and a media room, and an exercise area, the house becomes larger. Some people ask why it's so large, and I say, "It's large enough to hold two entire lives, and each with a part that accommodate every aspect of the two lives." The house is really an extension of both their personalities, together. That was one of the ideas—how do you represent that? That was another area where the form of the house could be generated.

When it came to bathrooms and the kitchen, they saw them as a necessity, but Bob and Eileen wanted them to be environmental spaces—not-so-enclosed works of art.

As it went into construction, we redesigned almost every element—in fact, every element—of the house, because Bob and Eileen thought every square inch of the house could be a work of art or architecture: the staircase, fireplaces, the kitchen island, every single bathroom. That idea extended into the spatial qualities of all the rooms. As an architect, I was trying to get my mind around this. Bob was against having rooms for the kids that you'd close off. He wanted to use the whole house.

We began to design with this huge pile of program, a virtual mound of material. We weren't trying to organize it all, but just trying to get it out there. I started to give it size, a location, as in a regular house. It became obvious that you had to use the entire 35-foot height that was allowable, because there was a building envelope in the site description within which you could build a house. The house did pick where it would go. The boundaries were irregular, since the site was irregular. That was work done by Gordon Winchell, and the architects who developed that property. So we had discussions with them, and the neighbors, and everyone was hopeful about what the house would be. Winchell was all for it, whatever they wanted to build. The other people, including the adjacent neighbors, were a little more skeptical about what would come out of it.

So when we began designing, after the programmatic phases, we ended up having five to seven alternative configurations for the house. Bob and Eileen and I would have meetings every week to ten days, and the discussions always centered around aesthetics, and never about the practicalities and functions, which were simple. They were more interested in the discussion. They were confident that things like the elevator, laundry rooms, and mechanical rooms would take their place in the building in the right way. So they had left it to the architects to make sure the house functioned. They weren't going to be functional watchdogs, and when the function got in the way, they opted for aesthetics.

Planning Board Minutes, February 16, 2000

Continuation of Public Hearing, Davoli Site Plan

The hearing was opened at 8:20 P.M. The proponents presented the issues they heard and their response to the last hearing on January 19, 2000.

1. Tree screening and replanting.
   - Met with the neighbors regarding the number, location, size, and species;
   - Revised site plan shows 88 new trees, 56 new shrubs, located and sized for natural appearance and optimal screening.
2. Visual impact from the pond
   - Moved house 10 feet back from the pond (any further impacts neighbors and water easement location),
   - Added 4 new deciduous trees between the pond trail and the house.
3. Two adjacent lots owned by Mr. Davoli
   - Within a cluster requires Planning Board approval;
   - Non-conforming requires Zoning Board of Appeals approval.
4. Lighting
   - No direct exterior lighting of house or glare from exterior fixtures;
   - Four new trees on pond side of house to further screen windows at night;
   - Farrar Pond properties are not required to provide a view of darkness to the pond or neighboring properties;
5. Building height calculation
   - Building Inspector reviewed the average grade and height, and concluded that the design is compliant.

Mr. Stimson, landscape architect for the project, said they have begun to collaborate with the McCann's landscape architect. The proponent has agreed to plant several trees along the path easement which will effectively screen the first floor now, and the second floor in a few years. The work has preserved water views for the McCanns. They have tried to address the neighbor and public issues. Mr. McCann said they have drafted a memo of their concerns. They also said that any movement of the house should be southwest, as moving it straight back could have a negative impact.

Members of the Board said they were impressed with the collaborative work, but continue to have reservations about the relationship of this house to the public domain, and suggested that the additional lot(s) could be joined, so that it becomes one parcel. Mr. Davoli said he would entertain limiting the use or intensification on the other lots. The other lots are independent. Their intention is to leave the non-conforming lot as is and use the other lot for a barn. There was a suggestion of moving the house further back. He believes they have compromised by pushing the house back 10'. Mr. Schwartz said pushing the house back further would mean a re-design of the house due to its unique use of the topography (underground garage is within an existing "bowl"). The Board continued to be concerned about the view from the pond. The proponents suggested adding landscape screening close to the source of the view. The Chairman said they are looking for a solution for both the town and the homeowner. The proponents suggested new trees to shield the house and break up the mass. Sidney Moss spoke in favor of the project. Mr. Cooper complimented the proponents on their work towards meeting the intent of the bylaw. He noted that the house was designed for a very specific location. He noted that they have used dark natural materials and natural landscaping. They have not made the house smaller, he added. Dr. Gordon Winchell, a neighbor, said he thought the impact will not be that great. Mr. Davoli said he is not prepared to move the house more or reduce the size. They are willing to put in evergreens rather than deciduous near the house, laurel bushes near the path. The chairman noted that the Board is concerned with precedent, with this becoming the leading edge of change in the character of the area; the Board is concerned with impact on the pond. The Board noted that building envelopes were originally thought as an area within which one might locate the structures for the house, never expecting a single structure to fill the building envelope. Would the owner consider reducing the envelope on the other parcel? Mr. Schwartz said the house may be precedent setting, but it may also be considered a model, in the way they worked with the colors and contours of the landscape as well as the re-planting of trees. The Board asked the proponent to calculate the size of the building envelope and the percentage of the building envelope which was taken up by the structure. A motion was made and seconded to continue the hearing to March 1, 2000. The motion passed unanimously.

Planning Board Minutes, March 1, 2000

Continuation of Site Plan Review/Public hearing, Davoli Property

Appearing with Mr. Davoli was Mr. Schwartz, architect. They have gone around the pond, and noted that the trail does not go completely around the pond, and due to dense vegetation they were not able to get a clear photo shot. They noted the house does not face the pond directly, but rather an inlet. He said the Davoli's will see other houses, more than they will see theirs. Mr. Schwartz said the building footprint is 5,697 square feet (including the overhang), while the building envelope is 10,220 square feet. In this cluster, the building envelope is the smallest percentage of all the lots, he added. He also noted that the design of the house is very site specific, using the "bowl" present on the land to create the underground level. Mr. Davoli spoke of the "confluence of many elements" to create this house on this site. The site will remain forested and the house will be a color that will blend in with the natural environment. Working with the neighbors and others, they will plant $150,000 worth of new plantings and move the house back ten feet in order to accommodate their wishes. He noted that Lincoln has a history of "embracing avant garde architecture." In discussing the size of the buildings, the Board noted that other large buildings, such as the condominiums, house many more people. Discussion turned to Mr. Davoli volunteering to reduce the size of the building envelope on the adjacent lot, or perhaps reducing the size of the house(s) to 4000 square feet. Mr. Davoli said he can't build on the old Fuller lot, and would restrict the size of the structure on the third lot to 4000 square feet. Mr. Breslin requested that the conservation trail easement stay where it is.

A Motion was made and seconded to continue the hearing to April 5, 2000. The Motion passed unanimously. The meeting adjourned at 9:45 P.M.

14 north elevation seen from farrar pond

15 main entry (south) elevation

design

Warren Schwartz, *continued from page 37*

Eileen loved the way the house would reveal nature, and she was interested in having nothing between the front door and the landscape. Discussions about views from the front door led to the idea of views from every room in the house, and that became a mantra: not just the view of the pond, but seeing different aspects of the site and the weather from every room. They were interested in all the views, some facing the woods, the entrance area, the tops of the trees, even the ground views. So you needed windows in every direction, all 360 degrees.

Bob was interested in having the house unfold. As you walk around, the house looks different from different points of view. You can't tell what one side will look like from the other sides. At one point, we were looking at a foundation, and Bob said, "That looks like a sculpture. I didn't know a house could be a sculpture." He was coming from a place where everyone else comes from. We shaped the house to follow its contours. We laid the house on the ground pretty much the way it was, and the house began to configure itself. Eileen liked that; she didn't want to mess with Mother Nature.

The site had a couple of major bowls. The garage, with three cars, was such a large element, the biggest thing in the house, and the advantage to these large kettles was that they could hold parts of the house. The kettle nearest the pond could hold the garage, and it could be reached with one small drive. So the house was built into the kettle, and it rises on top of that.

One of the things that made the design happen was just standing on the site, looking onto the pond, with our backs to the sun in the south. We oriented the point of view so you would be facing north, to avoid having the sun in your face, but the view would be illuminated all day, and it'd be constantly changing. Because the sun was at your back, you could bring the sun in over your shoulder, and warm the house in winter and summer through windows that were placed high. You'd see the sky but also have a lot of wall space, at human height, and above that, you would let the sun in.

About three of us were working on it in the office. We looked at the house being faceted on the outside and curved on the inside, and we made models and plans, for five or six different developments. But at one point in the middle, it became important to Bob that the curves be seen on the outside, and the faceted planes on the inside, and that the house would be flipped inside out. The challenge was how to make a piece of architecture that is coherent, with both straight and curved lines, in an almost equal balance, because the exterior is more assertive when you enter; but on the inside, many of the planes and walls are actually straight, not curved. So almost everywhere, flat planes play against curved, and vice versa, and that gives a liveliness to the house: how the planes catch the sun, and lead your eye. So their different takes led to this; it wasn't anything you'd know starting out, but that you'd learn in the process. We ended up having lots of straight lines, but also a lot of curved lines. Bob preferred the curves, and Eileen preferred the straight lines. One of the concepts has a design made only of facets, and the other, only of curves; and those two early designs merged, so the house was made up of both. The outside of the house was originally made up of flat lines, and inside, solely of curves. So I became interested in turning the whole thing inside out. It was a revelation to flip the house, not left to right or top to bottom, but inside out. The house became in my mind a big rope with a loop, one end sitting above the other end. The idea of the loop was that it captured the exterior, opened toward the pond, and lapped over the other part of the loop. Metaphorically, it was a natural, large, woven, nautical rope that might have been left in the woods.

And there was one other element about the exterior: their interest in stone. They showed me a stone house nearby, a light-colored stone. And after a while, we showed them stone in different colors, and tried to recommend stone that was more like the color of the sculptures they'd collected, but also like the bark on the trees that surrounded the house. Later on, when the town had to approve everything, and when it came to the question of the house's color, they were interested in a house that blended in with the trees, which they were required to leave, and it was more the color of the bark of the trees. But the viewers were skeptical and didn't believe that we could do that, because none of the other houses had been built in a dark stone. Bob was very insistent that they understand it, and that the house would be different from the preceding ones, and that he was different from the people who had built in Lincoln earlier.

But if their house was going to be stone, he wanted the house to be futuristic, and he also liked metal and glass. They liked the sheen of aluminum.

It'd be difficult to put all this together as curves alone, or straight lines alone, but I felt they wanted to make their house as full as their lives are, not one-dimensional. Most people don't have the luxury of doing that, but they really wanted to avoid the conventional house and start out here with a clean slate, as if, subconsciously, the house would be representing their life.

One question was how close to build to the construction line at the edge of the hill. The Planning Board asked us to move the house ten feet, so that it would be less visible, and so that any light at night would be less visible. So we moved the house ten feet back. Of course, ten feet in a landscape is almost insignificant, the size of a boulder. So we were happy to accept moving the house back that much. But we wanted to look out of the house and be able to see over the edge, because if you're back too far, you feel land in front. When you're closer to the edge of the hill, the more you feel the water, as though you're above the water.

The whole design process became like the music they like, jazz. Jazz has a structure, but it always comes out differently, depending on who the musicians are. We talked about the improvisational aspect of the house, holding it together aesthetically, while keeping it to the edge of coming apart. The challenge was to see how much you could put into a house, how much you could bring, and how much it could express. I think buildings and houses can do that, but not if they're bound by the strict rules of architecture. There are rules, but there are places where the rules can be kept in the background. Here, there was a lot more when the architect and the client laid aside their expectations for the time being, and just followed their own conscious and subconscious musings on what the house could be. But then, of course, it had to be pulled together.

The house was always on the brink of coming apart, and we were constantly monitoring how much it was coming apart as a design, and coming together. That was also part of the fun and the challenge: how close to the edge we could get without having it tip over.

So it was a balance. And actually, we didn't want either side of the equation to take over. We always felt, I always felt, at the edge of the whole thing, sort of hanging together and coming apart. It was important to do that. Bob was asking us to go out to the edge, not over it. How to bring it to the edge, bring in different materials to have a lot of them, and how do you sort it all out, and what about all the artists and architects that were contributors, and the fact that Bob and Eileen are different themselves and that their differences

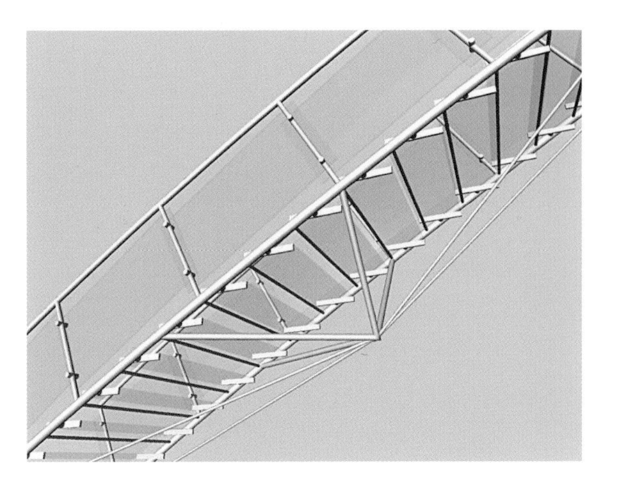

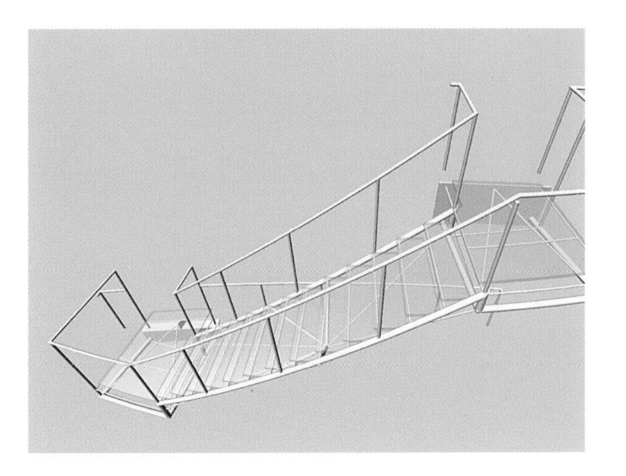

16 stainless steel and glass stair to third level

17 cast concrete kitchen island

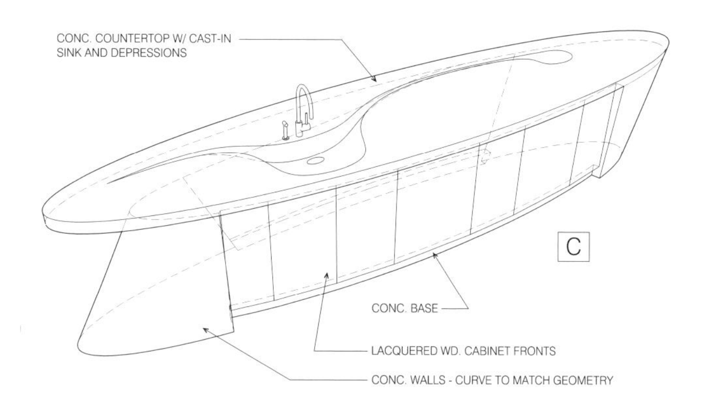

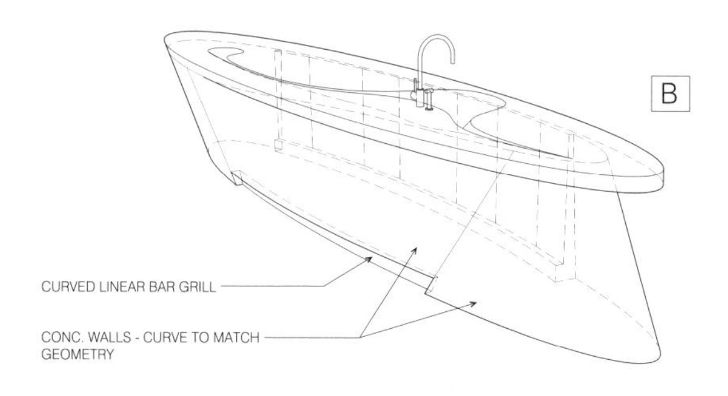

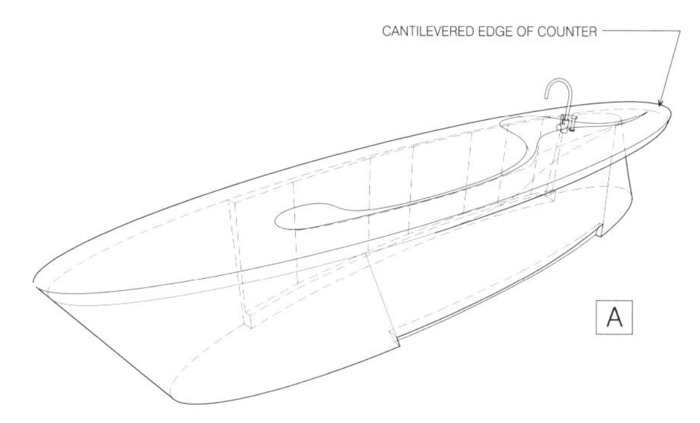

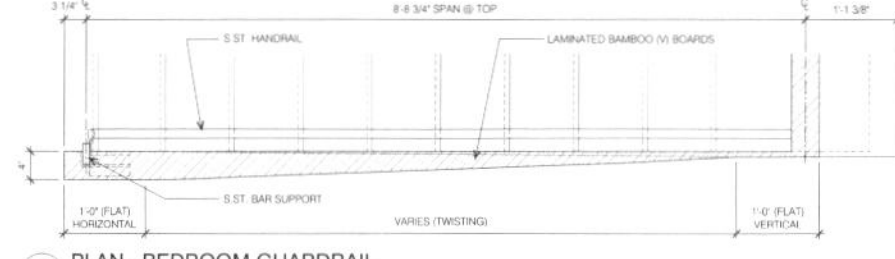

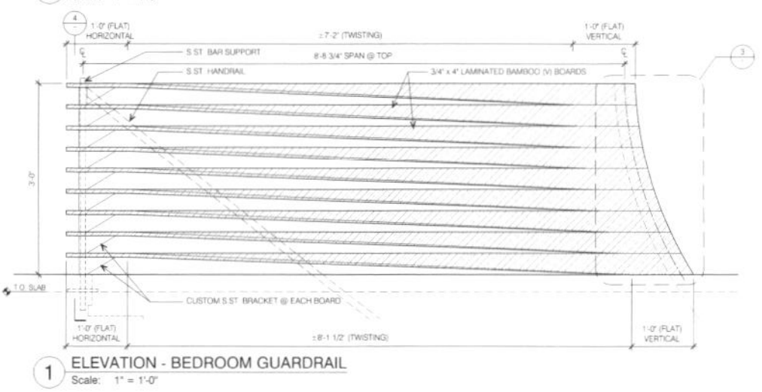

16 laminated bamboo guardrail and wall at master bedroom

18 diamond sawn black slate at Bob's bathroom

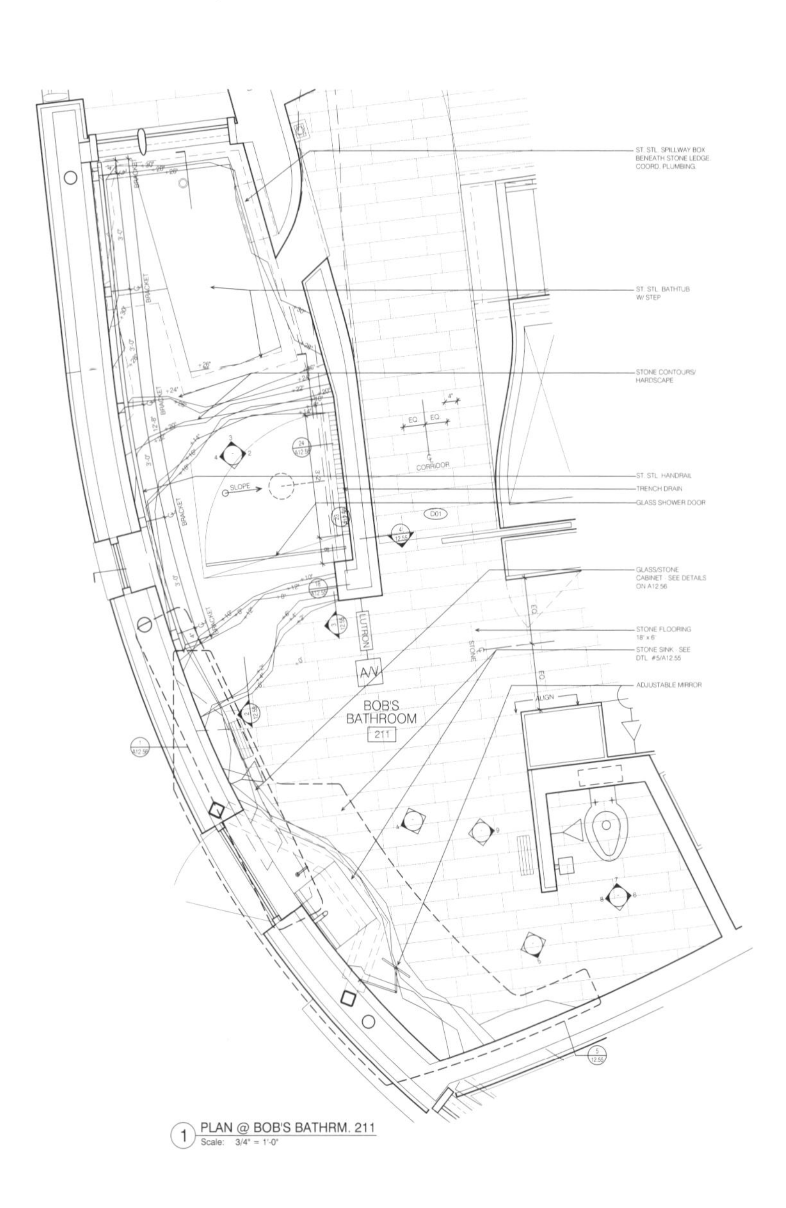

19 aniline dyed perforated wood panels at media room

20 living / dining areas at main floor

21 fiberglass wall 'leaves' at lower level bath

could be expressed in the house?

So it was a tall order, but it was an opportunity for discovery. We piled in everything that we could, because we didn't want to miss anything, and then the question was: Was it coherent? Could the result be understood materially, conceptually, aesthetically? Could it exist in a complex relationship of its parts, like two people's lives? Or during the design process, like three or five people's lives?

I enjoyed the fact that they were really becoming part of the design process, and they were suggesting all kinds of things. This lasted throughout the entire design process and even the building process, and we could discuss all things in a functional or aesthetic way. It wasn't the architect working for the client or, as in some cases, the client acquiescing to the architect. Here we were equal, both working for the house, as an extension of who they were and who we were, as well.

That was the most exciting part of the venture, not knowing where it would lead, but having faith anyway that it was going to lead somewhere. Still, I got this incredible pit in my stomach when I saw the aluminum leaf on the belly of the living room ceiling, because it's important to be at that edge, not knowing what you'll think of it, not knowing what others will think, but doing your best to keep it from coming apart. But we'd get weary sometimes, and then we'd say, "This is an opportunity. It's not anything that should make us weary. It should make us more alive and excited." It was all about that in a sense. I've always said to my partner, Robert, that architecture is like life, and he'd say, "That doesn't mean anything." I'd try to explain it: If you can do architecture, and make architecture, and make it embody the true complexities and complications of life, then maybe it will reflect it in some way. So it's not a style; it's something else, a record of a certain conversation with others. It's important to have the house exist in a realm where it's also not outside everyone's experience, but rather, it's out there on everyone's landscape. So that coming together and coming apart is an interesting explanation of what architecture might be.

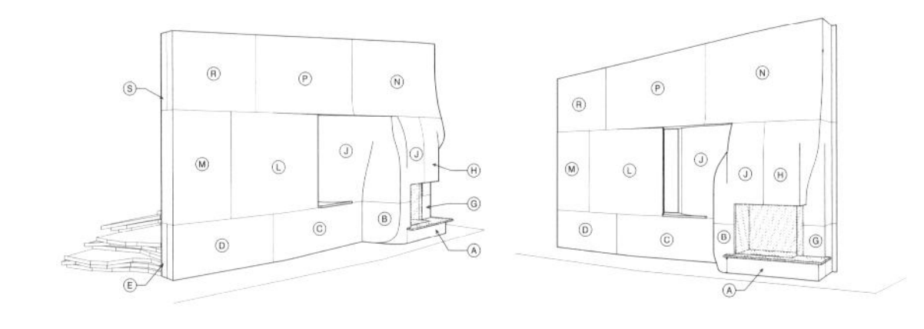

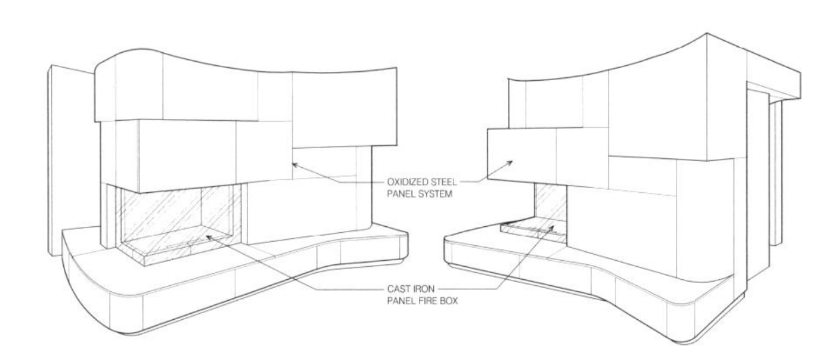

22 fireplaces

Planning Board Minutes, April 5, 2000

Continuation of Site Plan Review (McDonagh-Davoli)

Eileen McDonagh and Robert Davoli noted that the latest site plan indicates the removal of plantings which were to have taken place in a wetland area. Discussion turned to restrictions on the other two lots. While the applicants may plan a barn on one lot, they would like to keep their options open. They agreed to limit any structure on that lot to 4000 s.f. They do not wish to combine the parcels into one lot. Mr. Davoli said that his final proposal is to limit the size on structure on both adjacent lots to 4,000 s.f. The McCanns, direct abutters, asked if the Board had considered the conditions they were requesting, noting that their concern was compliance. The Board answered that they had read their letter. A Motion was made and seconded to continue the hearing to April 26, 2000 at 8 P.M.

Planning Board Minutes, April 26, 2000

Site Plan Review Continuance/Robert Davoli and Eileen McDonagh (continuation of a Public Hearing)

The owners presented a letter agreeing to limit development on each of the two adjacent lots to 4000 s.f. . . .While members of the Board continued to be concerned with the size of the house, the Board is also aware that this house set a standard in terms of design, materials, and landscaping features. Following further discussion, Ms. McDonough offered that while the structures on their other lot would be limited to 4000 s.f., the lot on Huntley Lane would be restricted to 3500 s.f. It was noted that the "gross square footage" would need to be defined.

MOVED

To approve the letter dated 4/20/00 with the following restrictions:

1. That square footage on Lot 1.01-8 will not exceed 3500 sf.
2. That arrangements satisfactory to the Planning Board will be made prior to any approval to insure that cost of installation of extending the existing water line will be borne by parties other than the Town of Lincoln;
3. That the Planning Board will engage, at the applicant's expense, counsel to advise the Board in the preparation of appropriate documentation;
4. Final planning Board approval will occur only if, and when, it approves such documentation.
5. The MOTION passed 5-0.

construction

eileen mcdonagh Feminist, Professor, Author

Personally, I would have been very happy if we'd never built a house on the lot. I fell in love with the land and the trees. Every time we cut down a tree, I nearly cried. We didn't level anything on the site, and we put in more trees than were originally there, but no house could be as beautiful as the site was without the house.

I don't like houses, generally. I like nature and mountains. I like to hike around a bend and see a new vista, and if it's in the Sierras, because they are high, you encounter different landscapes, and if you're high enough, the landscapes are lunar. With the sun hitting different peaks, the views are extraordinary. And as you hike up, it's different when you get to look down. What you saw as a pond or lake below becomes a shape. In your experience, you put together multiple views in you mind.

Bob and I are a good team. He's not as interested in the natural setting. He didn't want to buy the lot. He thought it was nice, but he's not as oriented to nature as I am. He's very oriented to building, and so we worked out a balance. I wanted to get the lot, because if we didn't, someone else would build a house, so we couldn't and it'd be gone. I told Bob that someone could always purchase the land from us, so we wouldn't lose money if we never built.

I don't like houses, because I grew up in Southern California on the Palos Verdes Peninsula, when there was no freeway and no high school—just open lots all over the place. As a kid, I loved to walk outside, and walk to the ocean. Geologically, it's an interesting area, with different formations, and a lot of places to take walks. But I saw houses gobble up the lots, and I grew to hate houses. I really disliked them. Because it was a zero-sum game; either the land was free and full of trees, or you had to find a path to walk around the house.

In Lincoln, we had the property for over twelve years, and I had no interest in building. We used to walk around the lot; it adjoined the water, which we loved. Farrar Pond is not as well known as Walden Pond, but they're similar—large, and you can swim and boat (as long as it's not a motorboat). I was extremely happy doing that, but Bob really did want to build a house. We would be doing the very thing I hated in Southern California—turning land into a house. Houses have stood for everything I hated since I was ten or eleven years old.

My parents had a tiny little view of the ocean through a large picture window, and you could see the whales go by and see them spout. I went to the beach almost every day. The rocks were very interesting. And through the window, we could see how the ocean changes with the weather. You got the effect of the weather all the time, regardless of what it was. There was always this sense of change, the idea that things are not fixed. I came to think that movement is important, that it signifies energy and life.

I had a biology teacher in college, and I learned it was difficult to define life from a biological point of view. At the time,

I learned that life is defined by the principle of motion occurring within an entity made of organic material. Animation within the entity. But it's difficult to locate life, the animation: Is it within the cell? Or do cells group to form a multi-cellular organization, the real entity where movement occurs? Does death occur when the brain cells die? Or the lung cells? It's an arbitrary definition. There are so many different cells. At what point is the person dead? If you can't definitely locate life, you can't definitely locate death.

If my first choice was to build no house, the second choice was to build one we'd like. To me, there was a connection to biology. Houses themselves are inert. People are alive in a house, but the house is dead. What I like about this house is that it's alive, wherever you are in the house. With all the glass, it's a composition that combines and synthesizes the nature outside with the inside of the house. I feel this house is alive.

I think it all happened because our first choice, the architect, was based on our interest in working with someone who has his own creative reservoir. We wanted to work with somebody whom we respected and had a creative talent, not someone who would do a good job in a style we already liked. Warren is known for being very creative, and for working with people who like to be creative, neither side knowing what they're going to do in advance. That really set a premise for what everyone would do on this project. There was Warren, who would hold the center, and there would be concentric circles around Warren.

The landscape architect was an important choice. We were drawn to a person who has the expertise, but someone with creative ideas that we couldn't predict. We found a new landscape architect, Mikyoung Kim, and liked her background—she's been a pianist and a sculptor—and meeting her, we really felt this was someone who would look at something in a fresh way.

Now, I really like color, and I've never had anyone help me with colors. But taking a look at this house in Lincoln, I realized I couldn't conceptualize a color scheme that would go through the whole house. I'd never faced that feeling before, so we had to look for someone who'd do colors. And because of the way the house was turning out, we probably couldn't go to a store and buy furniture that would fit into the architectural construction of the house. We were looking for somebody creative who could do colors and furniture. We found Calvin Tsao, who has a sense of color I've never encountered before. And you could tell that he and his partner, Zack McKown, would enjoy the project and be creative, and not just pull out something they'd already done. That would be creative even at the scale of the furniture.

And we had a previous relationship with people who'd worked on Bob's office, Monica Ponce de Leon and Nader Tehrani. We really liked them and asked them to do some of the built-in furniture and office areas.

So we got the feeling that we were forming a community of extraordinarily nice and gifted people. We liked that sense of collaborating among friends. So with the project growing, we expanded the group and invited two friends who are artists. At the time, I had this appointment at the Radcliffe Institute, and at a reception there, I met Shellburne Thurber, whose work had been featured at an Institute of Contemporary Art (ICA) show, entitled "The Death of a House."

She had photographed houses that were abandoned by their occupants. But in the photographs, you felt their presence. So we thought, why not ask Shellburne if she'd be interested in photographing the birth of the house? I'm really intrigued by process, like rehearsals and writers' drafts. The unfinished stuff. We wanted a record of that, because once a house is done, you don't have that first draft. And the house would be interesting at certain early stages, maybe even striking, with the scaffolding and the geometries of the curves and the grids. You can't leave the house in that stage, but you can at least photograph it that way.

Our group just sort of grew out of this feeling that because it takes so many years, we could integrate friends who are in the arts into the process. As we went along, we invited people to join.

Nature conservation is very important in these parts, I actually wanted to have a barn built on an adjoining lot we also own. I was thinking of chickens and Icelandic horses and llamas. And Warren was going to design a barn, a small structure, that would preview the house. But the same people who hated the house hated the barn—they hated even the idea of a barn. "Why would you want a barn?" they asked. I told them I thought I might like to have a llama, a horse, and some poultry. They said that llamas spit. They were really hostile, really hostile to even the idea of a barn, not the design of a particular barn, but any and all barns. That really surprised me, and what I realized was that that Lincoln has some people who are very suburban, people who probably don't like farm animals and don't like modern architecture. But Lincoln is still zoned for farming. It wasn't as though I was going to have a herd of llamas, but the idea of even one llama was too much. I still feel that it's good that people are expressing themselves, but what they were expressing was so disappointing. I thought we were leaving suburban values in Belmont, but we weren't.

Across the pond, where there's an inlet, an elderly couple in their eighties, and another elderly man, also in his eighties, owned three lots between them. They wanted to sell the lots so that the funds would be part of their estates, but there was tremendous concern that the lots would be built on. The topography goes up when you're on the pond, so you'd see houses at the top of the ridge. We were approached to find out if we would buy the lots and not develop them. They'd be our lots, but we couldn't build houses on them. In fact, we bought the three lots at full price, agreeing not to build houses on them.

But somehow, the rumor circulated that we were going to start a llama farm because llamas are pack animals—they don't do well by themselves. So because we had thought of having one llama in a barn, some neighbors assumed there must be a herd coming. By now all this had gotten so distorted, and it was disappointing. To me, that was an indication that somehow there was a fear, or just a suspicion, of difference, because the house is different, or because we wanted to build a barn. The reactions surprised me, given what I thought Lincoln was as a community. My major sentiment is sadness, because I guess I think that unless there's a really a good reason to be negative, you should be positive. So it makes me feel bad when people are so fearful that they become negative.

Walden Pond is not far away from our pond, and when I read the book a long time ago when I was in California, I'd never seen the pond, and thought the idea of pond culture was strange. California is basically an irrigated desert, and inland water is very scarce. What Thoreau called a pond would have been a lake in Southern California. Thoreau is very good at making you feel the pond is part of his home, part of his life, however. In California, I used to understand getting in your boat and going off on the ocean. I understood the oceans. But ponds? I had no reference point.

I liked Thoreau's book a lot, though, and when I was a graduate student in Cambridge, at Harvard, doing my Ph.D. in government, I used to bicycle out on weekends in the summer and spring. And when reading about Thoreau's cabin, I was struck by one thing: he didn't have the space divided up. It was open space, without boundaries, one not-so-big room, and that really appealed to me. There are separate zones, but nothing cordoned out. That principle, not dividing up the space rigidly but leaving it an open floor plan, is definitely in our house, despite the size and the fact that the styles are so different than Thoreau's.

The other thing about the site is that the sun sets over the water, as in California, enough that you get the sense of the water itself. That's a nice, comforting feeling. I'd go and live in the West, but Bob has no interest. But the idea was to create that feeling here, to capture the sun and the water and let them in, but hopefully with attention to the eastern tradition, too. Maybe it was naïve, but I think we were also following in the experimental Gropius tradition in Lincoln, which I really associate with the East Coast.

Mark Whitehead: Town Planner

This is probably the biggest single-family house in Lincoln, and it's a one-bedroom. People always give me a funny look when I tell them that. In terms of unusual shapes, we have a lot of different types of homes in Lincoln, which is probably more of an architecture capital than any town around. We have a lot of different styles, so things like this are not unusual. As far as I know, we're one of only a few towns in Massachusetts with a site plan review for single family houses, and so there is a whole process for review. Lincoln is a rural oasis in the middle of suburbia, and the reason they have the site plan review is to retain the rural character. There were other homes coming into Lincoln that didn't seem appropriate for Lincoln's style, because of their overwhelming mass and scale, so-called Mc-Mansions that have huge façades. We look for an architectural design that breaks that up. This was a big home but it fit into its environment.

The house was permitted before my position was created. When I first came on board, they came in for the fence, and wanted to make changes, and it was a new landscape designer. They have this kind of unique fence that's out there. The story I was told by the GC was that they had two German shepherds, imported from Germany, and one kept escaping. The fence looks like something out of World War I trench warfare. It wasn't attractive to me, but the Board authorized it. I guess it goes with the house.

We've had a number of different styles of homes over the years, so Lincoln got used to seeing different architectural styles come in. We had all the deckhouses in the 1950s. One Planning Board member said, "I don't want to be the member of the Planning Board who denies the next Gropius house." That's owned by Historic New England, and it's very much on our radar screen. The Gropius House quite often is held up when we're dealing with a different style of house. In another town where I worked, the people wouldn't have known how to deal with it. Here, we get a variety.

It was under construction for five years, but there were other projects going in around it as well. People get notified about a hearing for nearby construction, and we have to deal with complaints regarding the construction noise from this project. It was one of those situations.

shellburne thurber **Photographer**

When Bob and Eileen approached me in 1999, I had a show up at the ICA that centered on images of decaying domestic dwellings in North Carolina. After they'd seen it, I got a funny call. Bob speaks very quickly, and he said, "Basically, this is the deal. If you could do that with abandoned buildings, I'd like to see what you could do with the birth of a building." I was skeptical, because how can you talk about human energy moving through a building that doesn't exist, that hasn't been lived in? What are you photographing? What was interesting about the homes in North Carolina and the Boston Athenaeum, a library in Back Bay that I photographed, was that they had a lot of history; a lot of life lived in them. The spaces reflect that in ways you can't see. You get a feeling about them, a feeling that's hard to pin down. But a building that's never had anyone live in it is different from one that's been lived in for decades.

So this was a home that was being built. So what? I thought it might help to look at some of my past work to make it clear why I was hesitating, why it might not be the right project for me. But it was a remarkable site, and I really liked Bob and Eileen so I thought, what the hell, let's give it a shot.

It took two or three years to realize what I was doing, but I was intrigued by a lot of things. Every step of the way, the landscape changed radically and the building seemed complete for what it was. I started thinking of it as site-specific sculpture, and I got very attached to each stage, thinking it was finished. Just throw a tarp over it. There was this pure, unadulterated joy in seeing this thing emerge. Also, it's a very complicated structure in terms of what everyone's bringing to it and how long it's taken. Bob and Eileen wanted it to be a statement: a statement about themselves, about their values, about how they feel about life in general. I know they were careful about who they picked to design it. They have a close relationship to Warren Schwartz. They hit it off immediately. But what they asked Warren to do was an enormous show of trust, in a way—to let this person build this statement about them and create this thing that they would probably spend the rest of their lives in. A labor of love, all around. I think the place is remarkable, but would probably live in something quite different. Spaces that we live in reflect who we are and how we want to project ourselves.

Both Bob and Eileen have created themselves out of whole cloth. They're both mavericks in their own professions, and I find it interesting that they've built, from scratch, the home that they're going to live in. Most people move into houses that have been lived in before. So it's no coincidence that they're going to be living in a house that they created since they, more so than many people, are self-created. They exemplify people who weren't handed what they do. In a way, in photographing their house, I felt like I was photographing them. This whole business of self-creation is interesting. We're all self-created to a certain extent. But some people are born into something and they're scripted for life. However, there are some people who jettison their past and march into a whole new arena. I know for a fact that Bob didn't come from a venture-capitalist family. He wasn't born into a family of finance. He came to this from Upper New York State, from Syracuse, I think, and from difficult beginnings. I believe he was poor. He didn't come from a situation where it was handed to him. He came to Boston and had a series of jobs. Worked as a cook. He made his money as a dot-commer. Eileen was a legal scholar, and came up with a theory of abortion used in the Supreme Court, a groundbreaking defense of abortion that was actually based on a woman's right not to be invaded. I think it was used in a couple of decisions—I can't remember. It's easier for women now, but for a long time, law was seen as a male-dominated domain. I think Eileen's been out there carving new turf, creating it as well as defending it. It takes a certain kind of courage and a certain kind of nimble, creative mind to come up with it in the first place.

In a way, you could see this house built in LA, which is an architectural frontier with an "anything goes" feel to it. So to build it in a very old-world, conservative area, it's also like carving new turf. Lincoln is very old Boston, very New England, liberal and conservative at the same time. But it's also the home of Thoreau, and that's interesting, too.

Bob and Eileen were terrific about giving me creative reign. Over the course of the work, I developed a sensitivity to photochemicals, so I had to get up and running digitally. But it took a while. They were very trusting, going blind about what I was doing. I was finally able to gain greater speed and regularity and get the photographs out, but there was a lot of trust on their part in the meantime. They didn't rush the architects or the artists and artisans, either. They were also in the process of making changes themselves. But when you look at the house, it's so beautifully crafted. For a tradesperson, it must be wonderful to get the green light to use wonderful materials and spend the time. It makes me come back to my house, which I love for its imperfections. So much of it is jerry-rigged.

I think Warren was very sensitive to the site. I think that's because of all the glass, and because of the way Warren has worked inside to outside, working with the flagstone. The house feels oddly light considering its size and weight, which is massive. It's almost as if the building has gently fallen over the site and carefully enveloped it. It's a huge structure. I don't feel it's changed the site's character but encompassed it, gracefully and carefully, and kept it intact. So I think it's always been about the site. Even though the structure makes a very strong statement, it doesn't get in the way, which surprises me, because there are a lot of details to the house. You don't feel the building imposing on your sense of site. It becomes very light, which I love. The space, regardless of which room in the house, is impacted by where the sun sits in the sky. There's so much glass, and the interior is accessible to the surroundings, and you always see the pond. You really get to see it in the winter. You're constantly aware of the outdoors, so you can't be unaware of the seasons and how they affect the photographs. The time of year always affects how I photograph, but that effect is amplified here.

I grew up in a Victorian house and then, in my teens, we moved into a Frank Lloyd Wright-Gropius inspired house; and on some level, given an opportunity to photograph this house, it seemed disconcertingly similar to what I'd grown up in, even if there were no swooping geometries in our old house. It felt very familiar. That sense of being outdoors but indoors at the same time. At first, as a teenager, it was very weird. You didn't feel protected from the elements or protected from the view. There was a certain lack of privacy. Suddenly, you found yourself on view, like in a department-store window. It took me a while to get used to it. But there's a certain lack of constriction that you feel when you're behind walls with easy access to the outside. After a while, I gravitated to places that are open and airy, and I have a strong feeling now for open and flowing space, big loft spaces. Even though my house now is small, it has a lot of windows.

I would never be able to live in Bob and Eileen's house. It would make me too nervous, too exposed. It's a lot of house to live in, but they're both big people in terms of their presence in the world and their energy, so I think they can fill it. The thing that's so odd about it…it must be like being a small person and giving birth to a person who's 6'5", a really big kid or a really big young adult with a strong personality. I think it's intriguing to live inside someone else's vision. I started shooting when they started clearing the site, taking down the trees. I'd go in once a month and photograph whatever was there. I think what I found interesting was that in the past, whenever I was photographing—whether it was the derelict houses in Carolina, the psychoanalysts' offices, or the abandoned mental hospitals—I was bumping up against people I didn't know, or getting to know them without knowing who they were, specifically. So a lot of the time, I felt I was channeling something that I couldn't clearly identify. With this building, there was no history and I could talk about all the things I talked about in my other photographs, only without the history. It became a site of pure projection. It was not about a specific house, per se. In terms of my own work, this may be the last kind of interior photography that I do because, after this, I won't have to do anything else. For the first time, it included all the issues I bring to my work without the interference of a history that any particular dwelling brings with it.

Without a history, it was a clean slate. And in a very funny way, as I've been looking back over the pictures, I find it's hard to talk about them in a detailed manner. The house allows me to talk about space and light in an abstracted way. In the pictures, it's hard to tell what's in front and in back. They're much denser, more layered, more complicated, and that's exciting for me. I was stuck on the idea of photographing the energy that's been created over time, but instead I ended up photographing something I'd projected into space, so it became my self-creation as an artist. I straddle the line between traditional and fine-art photography. I came into this world as a documentary photographer, with a fine-art background. My work's always been informed by a love of painting that goes way back, in the way that space can be defined in painting, but not in photography. As I think about it, the building and I were, in terms of the photography, co-creators. It was a collaboration. But I don't think you could say the photographs are a strict document of the building of the house, because they're not. In many cases, you wouldn't recognize that it's their house.

I found the different phases of construction maybe more interesting than the final product. I loved the beginning phases, when the skeleton went up and when they started putting in the nervous and circulatory systems, and layered the skin down. All the beginning stuff was fascinating. When it was finally up and running and complete, as fabulous as it is, that's when it became its own thing, and I didn't feel as much a part of it. There was a certain kind of intimacy at work with this. I know what the place looks like underneath all the surface stuff. But people coming to it for the first time won't know what's going on underneath it, like the people who have worked on it for the last five years. For the builders, there must be a kind of attachment, some difficulty in letting it go.

So it's presented me with a different set of challenges, which in itself has enabled my vocabulary to grow. I'm sure these pictures are different from anything else I've done before, not because it's a different kind of building, but because I got to see it soup to nuts. I had never considered construction to be something of photographic interest before. I was used to reacting to something that was already there, but here I was reacting to something that wasn't there, but slowly developing. In a way, it got me thinking about constructing my own work, but I don't know where that's going to go. I got excited about building stuff of my own—not becoming an architect, but building environments. I'm sure it's changed me. Every project does.

The photographs on the preceding three and following twelve pages are excerpted from a collection of over four hundred photographs by Shellburne Thurber called "Construction."

?
40

Tyvek
QUIKRETE
ACME

CAUTION

DO NOT Remove

Warren Schwartz, continued from page 52

When the construction began, the house was pretty completely drawn and three-dimensionally described by the computer. The house was so volumetric and complex, it was difficult to describe in conventional flat drawings.

Just in terms of the surveying that had to be done to excavate the house, everything was done by computer. They made the smallest footprint they could in the forest to place the house in, and then began to dig the foundation, to fifteen and twenty feet below the ground. Thoughtforms took our drawings and laid out the points in space vertically and horizontally, so the steel could frame up the larger boundaries, which would then be filled in. The house is quite exact, and I think it might have been done without a computer, but not as well.

Bob was signed onto doing the construction process carefully and well rather than fast, and then soon after we began building, he realized he didn't want any part of the house to be anything like he'd experienced before. So he asked us to redesign the kitchen island, the fireplace, his bathroom, her bathroom, almost everything in the interior of the house—in essence, holding the place for those functions, but giving them a second look.

We embarked on designing them again. The first time, we'd designed interiors that were more conventional, but he asked us to push the limits again, to discard any formal idea about what a staircase was, or a bathroom was; he really wanted us to begin anew. These elements became expressive in their own right, and what was great was that they were much larger than the furniture but smaller than the house that they fit into, and were an intermediate scale, so that gave a sense of scale to the house. These elements give you a gauge to measure the space by being the size they are, and it gives a way to step down the scale from the larger objects. Rather than just two scales, the furniture and walls, these larger elements are between the two and define the inside.

And since these bigger elements—the large doors, the stairways, the fireplaces, the koi pool—were reconsidered, what happened was that these pieces became expressive in their own right: different forms, different colors.

Bob and Eileen enjoyed the house from the beginning, even during construction. The first party took place in the building maybe a year after groundbreaking, before there were any wall or enclosures, or roof. A beautiful restaurant setting took place in this really dramatic construction site. These events took place regardless of the weather, and whether it was roofed or not, they were markers in time. There was a party when we were half done. The whole process was something they'd taken on as a life experience. It wasn't something that was going to be rushed, not an end in itself but a means to an end. The enjoyable part was doing it, the participation. There were engagements, weddings, birthday parties, and just dinner parties. Bob Gustin would have to stop the construction sometime Thursday or Friday, and bring in tablecloths and chairs and tables, lots of chairs, so the house became a public venue. I think they've had at least eight or ten parties, and then the fundraisers for the ICA.

Bob Gustin was the genius of the construction process. It all came together in his mind. I think he's brilliant. He originally raised German shepherds. I had never worked with Bob, and he was amazing with Bob Davoli, always telling him exactly what he needed to know, regardless of what Bob wanted to know. There's a lot of frustration with construction, and Bob Davoli, who was extremely calm over the last five years, had an amazing forbearance. Maybe he reached his frustration threshold once a year., or five times in five years. And so, for instance, when there could have been frustrations about schedule, and when questions arose, Bob Gustin told Bob Davoli that the schedule was the way it had to be, which means that the schedule was attenuated. I think Bob Davoli appreciated that. Gustin didn't bullshit him, and I also think he was looking out for Davoli's best interest when it came to quality, when it came to issues of price. Gustin would lay out all the alternatives. Gustin was incredibly clear about everything that was going on, and this made it work, because Bob felt that nothing was covered up, that everything was out there, to see, to make the decision on. With all the decisions to be made by Bob and Eileen, it was an amazing feat to keep it all under control.

Bob Davoli didn't complain about it. He realized that this was part of a construction process that wasn't normal. Because Bob never did anything in a normal way. The whole construction process was a celebration, and Bob Davoli enjoyed it the most. Toward the end, he wanted to get in the house, but I think the journey was as good or better than the destination. Bob and Eileen are both members of the slow food movement, and for me, it's almost kind of a metaphor for the mind-sets that they had when they built the house. They let the flavors flow over them, and waited between courses. They took pleasure in all the courses being served. They really did appreciate each part of the house—the poured foundations, the arrival and erecting of the steel—because it was amazingly dramatic. With the steel up, you couldn't go up to the top of the building until the floors were poured, and once poured, you could walk onto the platforms, which were essentially a disaster area—there were no walls, so there were these flying platforms of concrete, resting on a steel structure.

So a lot of other featured items required special subcontractors. All of the woodwork inside the house was done by Graham Grallert, and it's amazing woodwork. Tripyramid, who built the pyramids at the Louvre, did the glass stairways. The guy from South Carolina who put the aluminum leaf in the ceiling was the best that anyone could find. The guy from Stone Soup poured the black-concrete counter in the kitchen, and it looks like a sloop cutting through the house. Just pouring this amazing heavy object, making it rest lightly on the floor of the kitchen dividing it off, was a feat. The concrete was pigmented black and polished, and the form was somewhere between architecture and furniture.

People like Calvin and Zack were called in to furnish the house. Everything that wasn't nailed down, or cast in place, they were responsible for. The dining table looks a little like the backbone of a dinosaur. It's conceived in pieces and built in fiberglass, so that each could stand on its own. I felt that these were elements of the house, designed to amplify all the aesthetic features. I was interested in them pushing their work to the edge. The large sculpture at the entrance and the large fence had to be strong, because everything else was strong. It was important to let each piece speak as loudly as possible.

You might think this would all yield a cacophony. But I was struck by the thought that it was an architectural playground where all the art in the house, and the house itself, was fine with everything else, without being regulated. I don't remember Bob or Eileen rejecting anything, except for something not going far enough.

If there were any disasters during construction, large or small, I'm not aware of them. In one case, I remember installing the stair, and it was a few inches too long or high; it was hauled up and had to be adjusted, and there weren't many ways to figure out how to solve this problem, because it was all welded together. But in the end, of all the other houses we've designed, this was the one that was the most

ROSEBURG

complicated, but it was also the one with the least problems during construction. I don't recall whether it was because of Bob Gustin, or Michael and him working together, or whether it was the computer, or everyone had signed on to do this difficult house. It was a reach for everyone but it went smoothly. I know that Bob Gustin didn't go home early, or come in late. Everyone was totally committed.

At one point, Bob wanted to have a sliding-glass roof over the dining space, to make something that opened and closed, so that it could rain and snow in the house. We said, "Bob, you've asked us for all sorts of things, and we've tried to give them to you, but this would be better to give you in the next house." We could have designed it, even with the weather we have here, but we would have focused so much on opening the house to the sky, I felt we wouldn't get the rest of it done. So I asked him not to request that.

I think the house could have been pushed further. The house was being pushed aesthetically, but it could have been pushed technically. Still, I think the house had to strike a balance, and I think it holds together aesthetically. It holds together, but not in a stationary way. It's a dynamic balance between all the elements—holding together while in motion. I think one of the strongest elements of the house, in terms of its motion, but also relative to everything else, is the curved ceiling, which starts at the front door, and goes through the front of the house and ends over the koi pond. It's such a large element, and it reflects all the colors around it. Somehow, it becomes a large organizing element both for outside and inside. It puts the space in motion, and the space flows around in a loop toward a fireplace in the living room. From the frayed end of the loop, it becomes a floating belly, blimp-like. So I think there's an organizing element that's dynamic, and reflects and absorbs all the other colors and materials, and organizes all the elements that are not reflections of each other, but individual pieces in themselves. You could talk about one element without talking about others. The kitchen element would be an element distinct from the stairs or dining table. I think they all add to the complex spatial nature of the house.

the house

23 approach from the east

24 west façade

mikyoung kim **Landscape Architect**

When Bob and Eileen came to the studio, we talked about music. I was a concert pianist before I went to Harvard's grad school in landscape architecture. I had trained from a young age, and I have an undergraduate degree from a conservatory, but for many reasons, I moved in a different direction. Bob went through the material I sent him, and when we met, he asked who was my favorite composer. I said Bach. It was an incredibly intimate conversation, like a conversation with friends, as opposed to, "Let me ask what you are going to do for us." I'm always prepared to talk in those terms with a new client. But, wow, it was puzzling for me when they left, because it wasn't clear whether they were very interested in my work or were just talking. They hadn't asked the usual questions.

An hour afterward, he called and said they wanted to hire me. But, he said, they wanted a different kind of house, and were bringing together different kinds of minds to collaborate, to shape an open arena to see what would happen.

That was interesting to me. That's how I'd describe my process with Bob and Eileen, and with Warren. It was a longer design process than most residential projects are. It was also a lab to explore ideas in a free way, both with the architects and the clients. In the discussions that we had, we were always together, even though in residential projects, landscape is usually considered a sidebar, subsumed by the architecture within its aesthetic and budget. It amounts only to decorating the front. But in this project, it was clear that the site was an important place that they wanted to respect. Plus, the building was embedded in the landscape, so its siting was very much part of the ground.

So the project was set up for the landscape to be very special, with a voice that would interface with the building. We worked with Schwartz/Silver on the interface, on the concertina paving that touched the building and went into it, and on the ground material and the views they'd pull into the building. At the front-door area, there's a way in which we worked with Schwartz/Silver when they did a layout that really reads like a sloped surface, a series of stairs, so that when you step down, you really feel there's a transition from the exterior through an ambiguous area to the interior. The exterior is just sliding into the building through the plate glass.

We took the discussions all the way down to how the lighting was done—implanting lighting in the canopies of trees to create this rippled and shadowed light play. Lighting is often an afterthought. We wanted to create these subtle considerations of lighting that you don't see at all. We have placed very low-wattage lighting and small lights in the trees, for these shimmering conditions that allow you to navigate the landscape. In the dining area in the living room, when you look out into the pond, there were two elements to work out: how we'd thin the trees by thinning out leaves to create viewing corridors, but also to reduce the amount of light, so no glare would come into the house. We also did that in the entry area, up in the canopy of trees. There's not much standardized lighting in the driveway, only three bollards of light. They're in Cor-Ten steel, just as a safety measure. Rereading Thoreau, we came to appreciate this incredible place, and respect it through the site. We decided that instead of imposing plant materials, instead of imposing an outside language, we would seek a language that knitted into the landscape. But it was not a natural environment, copying nature, but more of an abstraction. When you look from the living room down the incline to the lake, there's a number of birches, and we decided to use the language of birches as a gateway into the site. With white bark, and exfoliating bark with flesh tones, we decided to create these groupings of birches, which establish lines or gateways, at the front of the driveway and at the parking area. That's one of the contributions of the site, creating an experience of thresholds that you pass through with this iconic tree.

Throughout the entire process, Bob said, "No grass, no lawn, nothing suburban." So we followed two strategies on the ground plane. We planted an erosion control fabric to keep the ground from eroding. We embedded some seeds in it, and so created an armature for the natural ground material to grow from. So when you come in the driveway and look to the left, there's a depressed area that's a natural kettle in the landscape. We planted some birch across the driveway at that time, but basically we encouraged the natural ground material to come up, and it did. It's a natural, lush material. And in the corral, we planted a series of trees that really like the acidic ground condition. We're mulching the ground there, but with pine needles, to create a plane with a reddish, ochre ground that works nicely with the trees and will resist ground materials, and allow people and the dog to run through the area. Native pines and oaks comprise the plant materials.

Adjacent to the house, we created a series of paths where we wove materials with moss and thyme and created a woven surface, with inch gaps between the stone. It creates a ground-plane fabric of green and gray that transforms over the season. Moss is a really intense green from early spring through early summer, and it grows through fall and winter. And that's another thing—rather than a strong color palette, through the seasons, we've woven subtle blues, purples, and whites, both in the trees and the ground planes, with bulbs and annuals and perennials. Some die back in the spring, some in the summer, and others in the fall, for an orchestration of color that is layered through the seasons. Particularly because Bob is a musician, it was easy to talk about how a series of planes and places and voices could be contrapuntal. Within a structure, different colors, heights, and textures would come up and take over. Usually we talk in-house about the complexities and the layers that get us into a certain place in a design, but Bob and Eileen wanted to know—they were fascinated with the process. Often, the language we used was metaphorical.

In addition to the landscaping, there were two architectonic interventions that we did. One was the fence. I hate to call it a fence. We worked on it as a piece that was in design for two years, to make it into an element that really registers the ground. At first, there were no contours. The area was completely flat, so we worked with the contours that were there previously. I went out there and sprayed the contours of a grading plan with a can of orange paint, and we went in and graded and enunciated certain areas that worked with the fence. The mounds weren't big. It was a subtle topography, at the high point of the site.

For the fence, we made a series of flexible wood models, and played with different alternatives. The fence started off as a flat piece, and became more volumetric as we worked. We measured the shoulder of Lesko, their dog, for the limit of the width, and his height for the height of the fence. I often work with the body as a way of understanding the limits of the project, and the limits here were the dog. So we worked into the computer a series of construction drawings, both in AutoCAD and Rhino.

The fence is a unit construction with five unit lengths of Cor-Ten bars. They're built in this flexible system, which can be contracted or expanded at pinpoints. So the piece has this volume—it's not just thin, but can expand and contract with the elevational condition, and it also curves around. When the parts, a prefabricated pin system, are brought to the site, they're bolted together and then brought to a certain area and placed. Conceptually, each piece is almost the same, or similar—and on site, it morphs to meet with

25 stair enclosure at main entry

26 koi pond, overlooking farrar pond

the ground, and is then welded into place. What the three-dimensional computer models gave us was the limit in this tectonic system, just how much we were allowed to curve. The second sculptural element was a stainless-steel koi pond, embedded in a semi-enclosed courtyard, with a series of stainless bridges. Eileen and Bob are interested in nurturing these fantastic fish, and the idea of color was the inspiration for the project. Rather than a naturalistic type of pool, I was interested in creating a surface that would celebrate the colors and reflect what was inside the pool. One of the programmatic issues was that there's a lot of wild life in the area, and it would be heartbreaking for Eileen to lose any of the fish. So we were trying to create a fence, but one that would become a series of gestures over the pond—laminated bridges that criss-cross over the fish, creating these micro-climates in which to hide. And there's a large piece of cast glass that's right off the breakfast-nook area and sits atop two intersecting bridges. You can step onto the transparent ground, and look at the fish beneath you.

I think these pieces—the Cor-Ten fence and the koi pond and the ground-plane stone material—are very closely aligned with the work that I've done over the last three years, the idea of stitching things into a place, and to work off a tectonic, cellular logic, a constructive logic that is repeatable in construction, and logical in terms of how it's put together, but with variability. It's not a grid. It has an organic quality to it. It's inspired by natural systems.

27 north facade

When you look at something cellular, it has a natural logic. Yet, when you look at a tree, its construction from leaf to leaf varies, and I think that's ultimately what my current interests are. With the fence, the koi pool, and the ground-plane stone material that's stitched into the moss and thyme, you can understand that there's a system set up and that it's not a rigid system, but something with an organic quality to it.

Bob and Eileen basically engineered a relationship not only between them and us, but also between us and the design team, that was incredibly inclusive. They created a non-threatening environment in which everyone worked together. Throughout the process, they had a series of get-togethers, both intimate and large, to celebrate milestones and bring people together. I feel some of my best work has come out in this project because of that.

28 cor-ten fence

30 view west from master bedroom

michael price Project Architect, Schwartz/Silver Architects

I'm thirty-seven years old. This is the most design-intensive project I've worked on. We initially designed it over two years, but we've also been designing and redesigning through five years of construction. The process of redesign happened incrementally. First, they didn't like the kitchen island. And then, the fireplaces. It wound up happening to the whole house, room by room. We'd be done redesigning something, and there'd be something else. They picked at one thing at a time, and by the time we finished we had redesigned ninety percent of the interior, which was an interesting way to go about it—there was never a moment when we just wiped the slate clean and started over. You're designing one piece, and you're reacting to a context that's in constant flux. Each new piece has to respond to the previous design changes, as well as the remnants of the original design.

So at some point during construction, the project became like a renovation. The process often reminded me of the Renaissance, when the design and building of architecture would happen more or less simultaneously. They'd start building something, without knowing exactly how it might impact other things during the length of construction. The design process would evolve because of changing circumstances. They would see how things were going, and they'd react to it.

Maybe partly because the clients were so involved, they learned a lot, and over the course of five or seven years, they became more sophisticated. So things that they liked, or wanted, five years ago, changed. Things were demolished. Most people who build houses would like to do this. They'd see something built and decide they want to change it. This happened twice after the walls were up. They expanded the screen porch, which required reconstruction of part of the foundation. It was partly to improve the proportions of the room, but also to have more glass facing the water.

Whenever we were confronted with these changes, we tried to think of them as opportunities, not to look at them as last-minute changes, but as design modifications that were always part of the original intent.

Another element they changed was the landscaping. We wound up redesigning most of it with a new landscape architect. Bob didn't feel the original landscape design was edgy enough, artistic enough—those are words he used. It was quiet, it was nice, but it was too nondescript. It was understated compared to the house, deliberately so, in an attempt to highlight the building. The landscape was also about blending in, and preserving the semi-rural, woodsy New England feel of the neighborhood. When Mikyoung came in, her design didn't stray too far from these core ideas, but she brought a more artistic expression to the landscape, which Bob and Eileen wanted. The largest example is the fence.

It's been a pretty amazing project, particularly for me at this point in my career, and I've appreciated the opportunity to be so creative and have a client that's supportive in the way they've been. It's been fantastic. I've grown as a designer throughout the process of building the house. But it's the smaller pieces that we spent a lot of time on that were particularly interesting and exciting to me—things like the fireplaces. Maybe their house starts to reflect that. It has a big idea, which is engaging the landscape, inside/outside, connecting with the setting. It also has the ideas of being modern and expressive and different, and tailored to the personalities of the client, and the designers, I guess, as well. Even though the house has an unusual footprint and the curve is everything, a lot of the inventiveness is in the pieces, like the stairs, the fireplaces, and the smaller architectural moments. Warren refers to them as the bridge elements, because they mediate between the scale of the furniture and the building as a whole.

The fireplaces were one of the first areas we redid inside the house. Although our original fireplace designs were modern, they weren't the strong sculptural elements they are now. They were simple, square openings set within walls of slate fieldstone, so they had a traditional, modern sensibility–fifties or sixties Modernism. Bob thought they were too traditional, too ordinary. He came at us with the challenge to design fireplaces that didn't look like fireplaces. He said, "Show me something that's out of the box." I cringe at that phrase, but we got the point.

Essentially, he was acting like a venture capitalist. He didn't dictate what the new fireplaces should end up looking like. He just knew what he didn't want them to look like. The redesign of all four fireplaces took place over the course of a year.

A fireplace has a symbolic role in the home, and they tend to have a certain look. We saw Bob's challenge as an opportunity to break from conventional aesthetics. The symbolic role is the same—a place for gathering, the heart of the home—but we tried to find a new aesthetic language to express it. They have a boldness and a minimalism to them, which might be part of their artistic expression.

The new fireplaces seem somewhat autonomous, like sculptures embedded within the house. They don't look heavy. They don't feel part of the structural aspect of the building, although they do have major structural components hidden inside them. That sense of independence allows them to be more individually expressive.

The concept behind the bamboo walls in the bedroom area was inspired by the overall concept for the house. We tried to use the cladding as a surface element that could run between spaces and interconnect them. We thought of the walls as wrapped surfaces, with the cladding delaminating at the stairs to create a unique moment. The individual bamboo planks free themselves from the wall and twist to create a louvered guardrail around the stair opening.

31 kitchen fireplace; acid-washed steel

32 bath - lower level; fiberglass wall "leaves"

Behind all that, there's this idea of moving and viewing, the fact that this surface runs from one space to the next. The bamboo cladding actually starts downstairs in the screen porch, and it wraps up the stairs into the bedroom, and then it wraps around into Bob's dressing room and closet, and it wraps around the stair opening. It keeps you oriented, and draws you from one space to the next, because you're moving against it.

As Bob and Eileen pushed us to reconsider the various components of the house, like the kitchen and fireplaces, the challenge was to create a strong character for each without compromising the harmony of the whole. So an interesting tension developed in the design between the individual expression of things and their integration with everything else in the house. It's been fun to walk that line, and I feel that tension lends a kind of artistic energy to the space.

They were living in a smaller, conventional house jam-packed with artwork they had collected, piled up in corners. It was very cluttered. Part of the concept of the new house was to create a more gracious environment to display their artwork and give it its due. And there is plenty of room for sculpture, but with all the glass, there's not much wall space for hanging paintings or flatwork, which has been a bit of a problem. Whenever a decision came up between having a solid wall or glass, they always chose glass. They always wanted to maximize the view of the landscape and the pond. So the artwork was considered, but in a general way. Individual pieces were trumped by the beautiful views out of the house and the sculptural quality of the house itself. During the process, you worry whether these different aspects—the art, the landscape, the building, even the furniture—will seem like they're competing for attention. At the Guggenheim Museum in New York, there's an infamous tension between the architecture and the art displayed within it. I was nervous that the various parts of the project would fight each other in a similar way. But now, with most of the pieces finally together, I'm happy to see them settle into an eclectic kind of harmony. And relieved.

Earl Midgley: Building Inspector, Town of Lincoln

All the drawings, all the papers, everything, was checked with the various boards, with the people in authority. Davoli had to go before the Planning Board because the house exceeded the 6,500-square-foot limit. It's 15,000 square feet. But it wasn't really a problem, because it's in a private area, in a secluded spot at the end of a road, above a pond; and when the foliage is out, it can't be seen. This will definitely be a showpiece, but especially when your landscaping starts coming in, you'll never see it, because it's off the beaten track.

Everybody in the whole area around here has a respect for nature. We have an efficient conservation department, and fifty percent of Lincoln is in a conservation or wetlands zone. The department keeps a network of trails all through the town, some of the trails crossing the site, in view of the house. The trails interconnect, and you could probably get into Sudbury if you wanted to.

I've been here in town for thirteen years, and construction on the house lasted four or five. You just have to understand that a project like this doesn't go as fast as another. It's intricate and detailed, and the ordinary carpenter doesn't have a place here. It's for craftsmen. You couldn't take an average Northeast carpenter, because he wouldn't have a feeling for it. You have to be somebody that has a love for this kind of work, to create what the architect and designers are trying for. It has to be a homogeneous group of people working together. Somebody off the street,—he wouldn't even have the right tools to work with. They had a carpentry machine shop set up on the job, to make whatever they had to. Most guys have a power saw and a little skill and knowledge. But to make that curved ceiling, every piece had to be fitted to the next. Each and every piece has to be fit individually, one at a time, and each and every piece is almost a labor of love.

The average builder couldn't do this. It's taken one of our bigger and better construction outfits, Thoughtforms, to do it. Five years. Most of the framing is steel stud. Very little wood. It has an elevator down to the basement where I understand there's an area that's almost a bowling area. Everything is unusual. He has a pizza oven that will cook a pizza in eighty-nine seconds. They've had the best of craftsmen. And many engineers. It'll be something that will last an exceptionally long time, everything being steel or plaster. There are so many engineers and architects on this project, it made my job easier. Everything was engineered before they put it together. If I found some difficulty, there was always an answer. Somebody had already looked at it and had all the calculations and drawings, and showed me the reasoning. I don't know who the architect is, but what this architect did was equally as good as the engineering. It was a really complete set of working drawings that a good contractor like Thoughtforms could build from.

It's all very flowing, with curved roofs, curved ceilings, all kinds of glass, like something out of The Jetsons. A lot of dark stone in some areas. Almost to the point of being morbid, it's so dark. But in other places, it's cheery, with so much light coming in, shining onto beautiful stainless stairways. You get reflections everywhere. There's live moss growing on ledges of slate in the walls, and there's a large covered deck area with two or three different levels, with a fish pond underneath, where you can actually walk out and see the fish. It's got to be difficult to furnish a house like this, but the Davolis came in with things that are appropriate for the type of house it is. Like I said, different. But elegant.

There is nothing like it in the area that I can think of. There's that house that was done in the late thirties or something: the Gropius House. I don't think too much about it. It's nothing but a square box as far as I'm concerned. You want to think that this is something unusual. But to me, I'm sorry—it's just another house.

I only saw Davoli on the job, but to meet the man, you wouldn't think he'd have this kind of taste. He seems like an average person. Maybe a little different. Wears a black hat and black clothes. You'd think he came from Amish country. That would be your first impression. But that's not the case. He's an investor. He invests in corporations that are not doing too well. And he builds them up, and he gets a commission. It takes a different type of person, and different kind of mind, to do this. He can look to the future and not to the past. The man entertains in higher circles.

Many of the neighbors have not been happy. It's taken too long. Five years of equipment, with trucks coming in and out of the staging area. The neighbors could not have a very good impression: "What is this big thing doing in our neighborhood?" It took a long time, but it'll give value to their homes.

bob gustin On-Site Construction Supervisor, Thoughtforms

Our company does a lot of different projects, all very high end. This was on a different level. We do a lot of houses, two or three a year, each with something special, like a bowling alley. The projects are all custom, and when people visit, they say, "Aw, man, did you see that railing!" But in this project, every single corner was like that—the cantilevers, everything, everywhere.

Sometimes, we do high-end residences that are very Federal, with gorgeous, nine-piece crown molding that runs through the house. This doesn't have that. The biggest thing about this building as far as uniqueness goes was not the finish, but the structure, the geometry, the lack of the normal control that you have when you're putting something up. On any structure, even one as big as the Empire State Building, there's a kind of grid line, a logic that's very easy to follow, some kind of centerline or symmetry around which it revolves. When you frame planes, there's a control line, and you pull everything off that. This design didn't have one. It was composed of a whole bunch of radiuses coming together. We had to establish our own control. We came up with a way of using several columns from which to triangulate. Symmetry is not big in the house. The window mullions move on their own—there's a logic there, but you have to look for it. The kitchen island and the kitchen cabinets have a different radius, and the way that they've laid tiles is on a completely different radius, too. It's a house that's not stating the obvious: there's a little riddle to everything going on. This was my first experience with CAD. On the computer, you could see the logic, but it worked in a fictitious way. When we got in the field, it was a whole different ballgame. On the screen, you could see getting the structure up, but with real concrete and steel, the tolerances of each were different, and they didn't give in to each other. The whole steel connection of the house was based on a lot of fixed moment connections, and it could stand on its own like a steel skeleton that didn't require a lot of plywood, working like a diaphragm. It was one big moment frame. After we got it up, it was on to the waterproofing, and tightening up the house. Keeping it from leaking for years to come was the challenge. That's where it became a trying project. But the overall structure was the biggest challenge.

Most of my guys are carpenters, but aside from setting the doors in wood jambs, there's not much wood in the house. Using concrete, steel, and conventional window systems and curtain walls from malls and high-rises, we pushed the envelope of what you could do with the building wall. Actually, that was true for all the materials. The guys were asked to do things with anodized aluminum, and a glass-fibered, reinforced concrete. Outside, there was the mottled green and purple structural slate and rubble stone, from Vermont. The lead-covered copper roof will eventually get a patina, and so will the Cor-Ten steel used in the retaining walls and the fence. When you're coming down the road, it's not some big, white, McMansion-esque thing. With the landscaping, the way it sits on the property and the incline down to Farrar Pond, it looks like it belongs there. And especially for Lincoln and the artsy people, this one is out there with the Gropius House. There are other towns where I couldn't see a house like that. In Newton, it wouldn't fit.

There's no conventional palette of materials that runs throughout the house, like a substratum. It doesn't have that consistent background. It's very idiosyncratic. Things go against the grain of what you look for. There's a different organization, because it's really based on all these separate projects. Like in a museum. You walk from one thing to another, and the special projects are like exhibits. The music room has a carbonite desk piece, and you walk from there to the fireplace, with its concrete tiles and acid-etched steel. And there's a custom door kind of hidden in the wall. In one bathroom, there's Brazilian slate. So each room has its own thing going. When the aluminum leaf went on that curved ceiling in the living room, and the dining-room table came in, it changed flavors all over again. It's kind of fun to see the whole place transform with the different layers.

Thoughtforms has our own carpenters, and most of the time, they're working with foundation guys and other subs. Carpenters are creative by nature, and one of the ways they judge what they do and how they're doing is production—and on a house like this, production can be slow. They want to see things done and completed, something tangible. The Empire State Building, after all, was built in a year, and though it's a hundred of the same floors, it's still impressive in its own way. In this house, you go two or three feet, and it's different. There's been a lot of waiting. Getting the glass-reinforced concrete from Canada took sixteen, eighteen weeks, and we had to wait for the templates for a lot of panels. By year two, there's usually still a good attitude, but it's much harder to keep the guys motivated over a longer period of time, achieving that sense of accomplishment. So for most of the guys, and for me, last August was year five, and I think they looked to me to keep the rhythm of the job going and keep the pace. Even at the last stages, I had to keep them motivated.

I guarantee five years is the appropriate amount of time necessary to build this house this way. We could have built it in three years, but with all the architectural and client changes, there were plans an inch thick, and none of them were used. The plans changed mid-stream. It took a long time to get things underway. And it was a challenge to get a whole array of subs that would build things like the giant koi pond with bridges out of stainless steel. There are not a lot of people out there to do that. To develop the kitchen island with Michael, the architect on the job, we found Stone Soup Concrete out in Northampton. They're kind of hippie-like. Could someone else have done it? I don't know. We found local people, and they did it right in a timely fashion.

At first, I wasn't sold on the mottled green and purple slate, but the materials are not jumping out as much as I thought they would on the site. I try to remember what it was like coming upon the house. Most people were just blown away. They didn't ever really get to a point where they liked it, per se. They were struck by how it was done, especially people who are in the trades. Then, finally, you get quite used to it—we did, anyway—and everything seemed to be matter-of-fact. For me, it's been consuming. It's been a challenge for me personally, but I am glad that it came to a close. At year five, I was ready to let go and purge everything out of my head, and move on. And I think Bob was ready to move in.

34 bob's bath; black slate

35 kitchen island towards main entry; cast concrete

36 view towards kitchen

38 bob's office; sculpture by tayor davis

harry irwin **Project Manager, Thoughtforms**

Typically, homeowners have a hard time visualizing structure, but here that was true for the men working. They didn't know where they were. It goes straight this way, and then bends this way. It took a longer time before you could get a feel for what room you were in. When we put framing up for the second floor, we didn't know where the bedroom and media room were.

When the structural guys were moving the steel around and putting up the pieces, these guys would glance at the drawings, and it was hard to know where the koi pond was. Is this inside the house or outside? The way it flows now, it's easy to visualize, but when there were only a couple of columns, you had no idea whether you were standing inside or out. That made it difficult. It required a different mentality. You usually visualize where you're going before you go there. With the stone in Bob's bath, a Brazilian slate they use for doing fireplaces, you tell them this goes on that, and this sticks out there but not there, and they just do what you tell them because they don't know what looks right.

We do a lot of reproductions of houses that are 100 to 150 years old. You're looking at an eyebrow dormer, you're trying to work with a convention, and people look at it—they have a gut feeling that this looks wrong, but are not sure why. Here, you have no idea what looks right or wrong—you're just dependent on the drawings. Matter of fact, we were getting ready to build that fireplace in Bob's office, and Bob said he wanted the architect to think out of the box, and they did. The masons had no idea what they were building. It was not as though they could understand it and just build. You usually tell a mason where the centerline is, but here the architects had to draw the skin full size and divide the drawings in one-foot sections. Eighteen different sections. The architect had to spoon-feed the masons to build it. Even the landscape architect, she was a little more creative than most I've worked with before. That fence she installed that was a work of art was a little nerve-racking.

I think the guys found it amusing and different, not boring, and they really resigned themselves to thinking, "I don't know what I'm doing, but get it done." They had no confidence knowing that what they were doing was right. The steel guy didn't know. Typically, with ninety-degree bends, you can understand the elevation and see it in space, and walk away and direct the cranes. He'd look at the drawings, and knew the general direction where the steel would go, but it was like walking in a dark room with a flashlight. You can only put the flashlight on one thing at a time— no way to see the whole room at once, with the furniture stacked all over. All your brain can remember is that one thing. In the trades, your drawings make sense, and you refer to them once or twice, and assemble it in your mind. Here, you had to do it one step at a time.

If the trade guys got lost and had to proceed on faith, I had to run on faith, too. They did the bids but they took a shot in the dark. They put prices on things and proceeded

39 view towards farrar pond

with an estimate, but they were just groping and didn't have hard numbers. They didn't know how much steel, and what it would cost. My job was keeping track of the budgets. I had to justify the change orders and look at the impact on the schedule. The house took four years to get designed, with all the changes, like for the kitchen and koi pond. Our two-year schedule got thrown out the window. Bob and Eileen were more concerned about the design and the quality rather than the schedule.

It's a good-sized job in a lot of ways. The early part of the job, in terms of digging the hole, getting a structure up, a lot of that stuff wasn't just ninety degrees—the foundations, the structural steel, and the roof itself were complex, including the wood-frame infill between the steel. There was a lot of the roof, which means expansion and contraction issues; and where the roof curved, work slowed down. At the end of the job, there were a lot of different finishing materials. Because of most of those things, it was the most complex house we've ever built. And the biggest one-bedroom.

But a lot of the stuff was just middle-level difficulty, like the electricians who just ran wires, and the audio and data systems, which were no more complex than other houses. The heating system was routine. The painter just put up the paint—it didn't matter how the roof sloped.

Bob and Eileen were really open, looking for a lot of feedback on the design. A fair amount of rich people looking to build big houses are younger, with kids' rooms and playrooms. Bob and Eileen were looking at it for events, because they're so active in the political and artist communities. Bob wants sit-down dinners for functions, with ten tables, ten people each.

Bob made high-level decisions along the way, but they were both removed from the building process. He delegated a lot of decisions to the architects, and I know that they had a lot of design meetings. It was more common to have the design meetings over there in Boston than as part of the weekly job meetings at the site. Eileen went three months without ever coming to the job. Bob was a little more involved; he'd come out on the weekends and walk about. But he rarely came to the job meetings. Other clients, with fifty meetings a year, they'd come to every one.

I'm not the biggest fan of modern architecture. My background is carpentry, and many craftsmen are into Victorian, where you notice the craft. But that said, I find the outside of the house very exciting—driving up on the approach, perched there on the hill, the swooping curve—and there are great positions in the house where you look to the outside and see back inside again. With all the glass, there are great surprises in the house overall, unexpected views.

That belly ceiling that goes from inside to outside–hundreds of people worked to make that happen in ways that are not visible: the structural guys, the metal framer, the plasterer who had to fare it out, and the painter that prepped it for the aluminum leaf. You get one solitary view of it, one impression just when you walk through, that starts with the front entry and goes through and out. When you see that, it gives you one sweeping feeling, and you don't realize it took hundreds of people to get it right, down to the aluminum leaf. All that had to happen to make that one big "wow."

Again, the architect gave us that drawing of one sweeping curve that goes outside and back inside again, like a billowing sheet, or a cloud, but I don't believe that when you look, you think that it was craftsmen's hands that made something like that. I was hired by Thoughtforms as a finish carpenter, and when you see old houses, with traditional, heavy crown moldings, you can imagine the person who makes them put them through shapers. It cries out that it's hand-done. With this modern architecture, the strokes are much broader, and the craftsman's hands are overpowered by the architecture, and you don't think about the guy who did the work. The slate? That's laid-up stone outside the building, rubble slate that's obviously laid one stone at a time, the same way it's been laid for millennia. There's a huge amount of handwork in the house, but a lot is hidden by the architecture. Maybe you see the craftsmanship on the handrail. With the twisting bamboo, you can visualize guys putting it together. It's simpler, it's a neat little idea, and it's still close enough to woodworking. But not the big-belly ceiling. Hard to see handwork there.

40 view towards dining/kitchen

tony davlin Inventor, Pizza Oven

I'm a freelance inventor and glass artist, and I've built a lot of ovens because of my glass background. Bob's a musician, and he visits Club Passim's, in Cambridge, a crunchy place where they have all this music. I think he's on the board—it's a non-profit—and they have a little restaurant there, and they have a pizza oven. The guy who runs the restaurant paid a lot of money for an oven that was a piece of junk, and he was in litigation with the manufacturer to get it to work. He'd paid over $40 thousand and couldn't get it to cook a pizza properly. A vegetarian chef there, DeeDee, knew my friend Steve Moskowitz, who contacted me to see if I could work on an oven like this. I got it to work, changed some of the systems, and Bob was getting some pizza there, and "Wow, I'd like to get a pizza oven for my house," he says.

Maybe I got it to work, but it was still a piece of junk. When I was studying the oven, I noticed a lot of shortcomings. It overshot a lot, and wouldn't keep its temperature. If you'd dial in 500, it'd go up to 600, then go down to 400. If you try to cook something, you want the same variable, and not the oven introducing other variables. Also, sometimes it burnt the pizza; other times it undercooked the pies.

Basically, the oven didn't work that well, but it cooked quickly. So I'm an inventor, and the first thing I did was to search the prior art, and I pulled up links, which go back patent to patent. So I came up with an idea to produce a new type of oven, and I made a proposal to Bob to build one for his house, and I made a dog-and-pony, eight-page product presentation, including my background. I brought a pizza stone along with the presentation, I explained the type of oven we'd like to build, and made a deal with him that if we made the oven for his new house and sold the technology to a company, we'd give Bob his money back, and he'd have the oven. If we license the technology, Bob gets paid per oven and gets his money over a period of time. We made it a business proposition. It was a short meeting, twenty minutes. The contract for a product is pretty simple, a three-page agreement. We signed it. I'm not a focused guy, and I've got other things to do, but at that time of my life, I was really into building a good pizza oven. We'd work with the architects.

So I met the architects about the oven, because just based on the house's design, the oven had to be round, or at least it had to have a lot of curves. A lot of problems with the oven were the same ones as the problems of the house. The oven's built like a tank, with an internal stainless-steel frame, and then we wrapped the inside with insulation, and wrapped the outside with stainless steel. It's like the house, which has a steel frame wrapped with insulation that's covered with plaster inside and masonry outside. There's another analogy between the two. They're both round, and it's much harder to build something that's round than square. Most houses and objects (i.e., ovens) are not curvilinear.

I met with the architects, and we talked about their concerns about how it'd integrate the oven with their kitchen. First, we made a stainless-steel part, the lips, which are going to be visible. They encased my oven except for the lips, and I went over their design. There were a bunch of mechanical issues to deal with, such as ventilation and power. Making the oven itself was quite difficult. I'd built ovens that melt glass at 2,000 degrees, but I'd never made anything for food. I needed someone to work on the oven with me, and my associate Eric Starosielski built it and added his equipment-building expertise. Eric and I come from a glass-blowing world, and it's not the safest thing.

We definitely over-designed it. The oven has a lot of complicated parts to route the air. We built it, and it wasn't getting hot enough, and we had to put in a bigger heating element. And then we had to rewire it so it didn't kill Bob when he's cooking. When you make something new, nothing works; it takes a couple times to get it to work. Eric's dad is an engineer, and we put fuses in the oven and built it to UL standards. Until then, the building inspector wouldn't sign off on it at the house.

We used the computer, designed it in CAD, but still it came down to filing things to get the parts to fit. Nothing worked perfectly. Eric built it. I'm the idea person but not that practical, and Eric put a lot of time getting it to make sure it didn't leak air. In some ways, it's like building a motorcycle, like on that show American Chopper, so it'd be safe. There were some parameters with the pizza oven that couldn't change, so there wasn't much interaction with him and the architect about that, except to just get the oven to fit next to the wine cellar and the Subzero.

Bob's hired so many people to work on the house: people designing the furniture, the landscape. That whole spirit of hiring so many people to work on it was really great. And then seeing him at these parties! He's into folk music. He met his wife babysitting for her boys, and then went to computer school, and made a lot of money, but he still has a lot of these interests.

I think the whole house is like the pizza oven. Everything is custom and different, and Bob pushed the architects. He wanted everything about the house to be special and different. There's the koi pond, with the life-support system, and Bob's commissioned a lot of custom furniture. The pizza oven is not an artistic thing, but a technical thing.

When you attend Bob's parties, you immediately know the house is different. It's not just a McMansion with crown molding and granite. Most people who build houses like this don't care so much about the process, but go with traditional things, like a Georgian house with a three-car garage. I believe Bob's on the board of the ICA, and he met the architects through that. The process with him is part of the deal, and he becomes part of the process.

We underbid the oven. I was sure we could build it for $40 thousand, but when things didn't work and didn't fit, it got more expensive, but that's the way things go. But I'm happy I built the oven. I learned a lot, and it's probably the nicest thing that Eric and I have built. We also learned a lot about designing something on CAD. Bob's had a couple of parties, and the oven works, and that's satisfying, and the chef said it's consistent. You put the pizza in and in a couple of minutes, it comes out and it's done. You had to baby-sit the original oven. It'd be really bad if we spent this money and it didn't work. Generally, we were pretty lucky with the oven, because it cooks pretty evenly. I'm happy it works so well.

Eric and I are trying to sell it, but if you have a new idea, it's tough to sell it to people with an existing old technology. It'd help if we had another prototype, which we need to make a gas oven. Next time, we'll do two ovens, so we'll have one to go out and sell. Bob's is the only one. We are currently in discussions with a large company about licensing and improving the technology.

41 master bedroom fireplace; glass fiber reinforced cement

graham grallert Custom Millwork, TFC Studios, A Subsidiary of Thoughtforms

Some of us have boat-building experience and we used similar processes in this house. But in other ways, this house was more complex. You have a certain amount of artistic license in building a boat. Here, we had starting and ending points that we had to adhere to. There was no place to hide a problem. It was constant problem-solving throughout, which I love to do. This was one of the most challenging projects I've been involved in. We were able to apply experience we'd had, but it was a learning experience for us all.

We began our custom-millwork shop fifteen years ago, when we found there was a need for filling in the spaces between contractors, when something unusual wouldn't fit the format, usually involving curves. This house is full of crazy curves, so it seemed a good fit to take on the house. In almost every way, they veered away from standard procedures. Some of the cabinetry was subcontracted out to other cabinetmakers, and those pieces were basically rectangular and could be manufactured by a more mechanized shop. This shop is all talented, hands-on craftsmen. Even the straightforward stuff was unusual, like the detailing and the materials used.

We did the primary cabinetry in the house, a large percentage of the bamboo walls, and the torqued, curved kitchen cabinets. We had to figure out how to fabricate all the curves, and there was no place for lack of precision. So there were a couple of places in the house that really challenged us, especially the railing that morphs into bamboo walls that sort of bump in and out, and then there were the cabinets in the kitchen.

Some of the bamboo walls are a variation of a ship-lapped wall, which is a standard sheathing procedure that's hundreds of years old. We use the same procedure on many of the inside walls, but only on the flat portions. On the curved portions, we couldn't apply the single boards.

How am I going to explain it without using my hands? It had to look the same as flat walls, with hundreds of horizontal boards, but it's one single curving plane with planks that have to align. We built up pieces of wood and epoxy, and pressed them together using a vacuum press, to create the contoured surface, similar to the process that the furniture-maker Eames used. We applied strips of veneer, a thick final veneer, one eight-inch-thick, to show the joints between them. The lines are perfectly horizontal but run through a curved surface, so if you were to lay each piece flat, they would not be straight-edged, because they have to travel over this organic surface. So the curved wall looks like a flat wall that was pushed in. Picture this: If you had a piece of rubber in a vertical plane with horizontal lines, and you pushed in with a sphere, all the lines that are pushed in stay horizontal. If you were to take that undulating surface and freeze it, each one is a subtle S curve, and different from the next. It must be nice and smooth throughout with a subtle, calm surface. If you walked up and it had lumps, it'd be a failure. It's a place where precision is critical, because of its simplicity.

ilan averbuch New York Sculptor

I'm involved in a lot of public projects, and usually I get on board at an early stage of the design. It doesn't happen every day that someone comes and asks you to be part of a project at an early stage for a private home. To me, it felt like an Old World commission, like in the Renaissance, when a painter would consider the direction of the light from the windows, and compose the light in the painting in the same angle. Nowadays, you usually exhibit in private galleries, where you make your art for a neutral "white cube." From there, it may end up at some people's private home or an institution, without being part of the planning of the artwork.

When I came for the first time to the Davoli/McDonagh property, I saw the forest around the house, and the trail diving down into the property, and then climbing up toward the house. There was a sense of driving up the hill to a temple or sanctuary. I am originally from Israel, and this kind of experience had many associations.

I wanted to have a staccato between entering the property and reaching the house. I wanted to have an event happening at that low point in the trail, prior to climbing up and reaching the house.

The result of my work is that when you enter the property, you're driving in a small, narrow lane in the middle of a manicured forest, and you discover a procession of abstract wood and stone figures next to the trail, crossing a monumental stone frame slightly tilted. When you arrive at the sculpture, you then capture a glimpse of the house. I wanted it to be like in a novel, where you have experience after experience leading to a climax: in this case, the mansion on top of a hill. The stone frame is framing the forest, and from some angles the house, too; it depends where you're standing in relation to the two. When I first came to visit the project, I immediately could sense the strength of what would be the house, although it was just a structure, without many of the walls and the details. I could feel what would be the experience of discovering all of a sudden the house on top of the hill.

When you approach the gate to the property, you do not see the house right away. As you cross the entry point, you are doing the same thing as the procession of "figures" in the sculpture are doing: crossing a threshold to another experience. In creating this image and placing it where I placed it, I was setting up a preliminary sensation that describes the experience you are about to go through.

The feeling I had with the Davolis was of people who are a combination of being invested in academia as well as invested in developing things in the real world. I had a feeling they both came through, in the whole picture of the project, with the academic's intellect and the investor's exuberance. They are a very unusual couple, and there was a lot to talk about beyond the intellectual dialogue and the client and artist conversation. Mr. Davoli plays guitar, and Ms. McDonagh is into horse races. I went to a mid-project a mid-project party at the house, and it was an interesting event, a group of very creative people sharing the occasion.

I met the architect Warren Schwartz when I first came to see the site. It was before the sheet-rock work, and some of the external walls were done so you could see the structure and the sculptural feeling in a lot of the elements of the house. He showed us the beginning of what would be a kitchen table: I thought to myself, "That could pass as an interesting sculpture." I remember looking at the roof of the first floor, from the top of the building, all covered in lead-coated copper. It was incredible to see that level of detail. Inside the house, there were many interesting fixtures, as well. Usually ceilings are flat, or in old Arabian houses in the Middle East, concave. But there, it was the reverse—the ceiling was convex. There was a strong feeling that you're under something, like the bottom of a boat. I could not say that the forms of the house directly affected my sculpture form, but it was nice to feel I was inside another "sculpture."

I enjoyed spending time in that part of the country. It's a very historical environment, seeing the pond, and thinking of Thoreau's famous text, Walden. Driving around and seeing the historical towns, it was like an American history book coming alive. Even the graveyards are incredible, full of famous names and historical figures.

 entry sculpture by ilan averbuch

My first experience with the area was with the De Cordova Museum. I have a strong presence in the museum, with two outdoor sculptures. It is just around the corner from the house. I don't know if Mr. Davoli saw my sculptures there or in some of my shows in NY or elsewhere.

I loved the whole experience, and it was nice to work next to an architect who does such accomplished work. I look at architecture for inspiration. Architects still trust form, while art these days is often conceptual and shy of material and form. Sometimes, when I go to a museum, I'm as much interested in the building as the content.

This house has a lot of interesting things going for it. It's edgy and constantly changing and shifting. You cannot rest—you have to move and change with it.

 entry sculpture detail

calvin tsao **Interior Design, Tsao & McKown Architects**

My partner Zack and I got a call out of the blue from Bob, who said that Liz Diller, who was designing the Institute of Contemporary Art in Boston at the time, had recommended us for the interior design. I said that we don't really do interior design, only the interiors of the buildings we design. But Bob said we're the best in the business, or something like that—I didn't know Liz regarded us that way. We deliberated and thought, "Why should we do this?" And then we realized the building was being done by Warren, and when we were at Harvard, we always felt that Warren's work was so progressive. Also, Bob and Eileen fascinated us. They make such an interesting and gracious couple; we sensed they're iconoclasts and were interested in the experience. We likened our design to a journey, like a trip with a group of people. I think the impact of interior design is strongest when clients are fully engaged in the process. When people start thinking of us as "well-known" architects and interior designers, they expect us to repeat ourselves. With Bob and Eileen, we realized, we could break any formula and work in totally unchartered territory.

This building has been designed to a fare-thee-well, every inch of it architected. Because of Bob and Eileen's infatuation with process, we were interested in being part of the team, part of an unfolding process. I love that they embrace everyone who worked on the project. I haven't experienced that since the sixties. It's charming and sincere.

We saw the interior as an interior scheme within the broader concept of a landscape, like a Chinese box, and conceived each room to be dominated by one piece of furniture. It wouldn't touch any part of the architecture, but simply float like a jewel in a box.

We created just four pieces of furniture. Besides a conference table, there are sofas, including the figure-eight couch. And, in the dining room, there's this big table that can break down, so that three, six, or fourteen people could dine. In a way, the furniture is grafted onto the architecture, not just sitting there but also intersecting it. Besides our own pieces, we helped them purchase the rest of the furniture. We formed part of a fraternity with Nader Tehrani and Monica Ponce de Leon of Office dA. They'd done Bob's offices, and here they did "his & hers" offices. We took on the more domestic rooms and the more public living area. There's a clear delineation between Bob and Eileen's working lives and domestic lives. Within the landscape, the building is a cluster of forms; similarly, we translated the juxtapositions of architectural forms within the house to furniture scale. There are relationships to the enclosure and its space, between the solids and the voids. The objects themselves, while performing as functional furniture, also have space, and a vibrating resonance between the building envelope and each piece within it.

What I love about this furniture is that it suggests a way of living—ways to sit, to lie down, to experience sensuality. Do you want to slouch, sit straight, or perch? In a way, it's an extension of how Bob and Eileen perceive their body image. There is a free-form couch, built with various densities of foam, that in a funny way, reflects how Bob and Eileen sit.

44 view towards living area/farrar pond

45 view towards open court / koi pond

Bob is almost always ready to get up. Eileen is very different, more settled. The furniture takes on their postures, their body rhythms and habits. The form becomes a result of how they sit together, often leaning on each other. It's a character portrait. It's personalized, anthropomorphic portraiture.

We always see the kitchen as a place through which you can come and go. It's a dynamic space, but you can also sit and think in a variety of ways—perching, leaning, sleeping, or sitting with or without coffee, next to the sink counter. And the kitchen responds, like a diptych. It's interactive and doesn't dictate a prescribed way to use it. The dining table is similarly versatile; you can take it apart, interlock it in different ways, and create a shape that's like a circle chasing its tail.

We went to the furniture fair in Milan, to get an idea of what's out there in the world of contemporary furniture. Bob and Eileen fell in love with the Campana furniture, so for the living room, we got two of those large piles of spaghetti. They totally expect friends to throw themselves on this furniture, like a super beanbag. Once a sixties person, always a sixties person.

Zack and I also determined colors for the space. Part of our exercise was to pull spaces together; for example, the colors of natural materials like slate and steel would be brought into some engagement through the intermediary of color. All materials were considered except sheet rock. Materials define spaces and generate volumes with their surfaces. All the other materials defined themselves, but the sheet rock needed that qualification, and that ended up being our job. We looked at it atmospherically.

Because the house is so transparent, the colors that predominate in the landscape come through, along with the colors of the house's material palette—the bamboo floor and the green slate. Somehow, that all added up to an atmosphere. We thought: How can we physicalize the atmosphere with colors that blend the manmade and the natural? The architectural material already reflected nature, and we extended the coloration. The common denominator is the greenery, but the other colors are smoke, the color of afterglow at sunset, the deep violet-blue that occurs before night, and the color of the sand that goes into the edge of a pond. It's like a wheat tea.

We worked with a Belgian paint company, using a tremendous amount of pigmentation. The colors are complex in a Josef Albers kind of way, one color shifting hue because it's next to another. We tried to play with colors to tie the furniture in with the architecture, but the biggest issue was to calibrate those colors for the ever-shifting light that's near a body of water. It's not iconic orange, red, and green, but colors that are natural conditions rather than pigments. And that shimmering surface, the ceiling with the aluminum leaf, has light bouncing off it and reflecting across the floors, which in turn become effervescent. There's something specular going on, an ephemeral effect with a vaporous quality. The volume is no longer a volume. It dissolves.

We always look for clients with whom we can be friends. You learn from the relationship. That connectivity is very important, since most of the time, we're locked in an office, working away. Bob and Eileen, as a couple, seem to be a paradox. Bob is hyper, but he's ultimately a child, as is Eileen, but she's also a center of gravity.

There are still missing pieces, but you know what? The house should never be finished—it should be dynamic. Bob and Eileen love to buy art; therefore, this is a house that needs to be curated. In the end, we may help them organize, but they'll make it their own, as they should. They've commissioned an artwork that has its own characteristics, its own presence; but their own lives, with their children, grandchild, cherished friends, and beloved dog, Lesko, will not conform to, or want to conform, to the construction that all this implies. It'll be interesting to see how the vectors interact and evolve. It's a playground, and it should accordingly remain playful and ready for change.

46 view towards dining / main entry / stair

47 media room - second level

48 bob's bookcase/desk; carbon-fiber

nader tehrani Boston Architect, Custom Fittings

This has to be one of the rare times that we feel the clients were completely embedded in the project. Bob is a very committed museum trustee—genuine, sincere and earnest—and he doesn't challenge a project to restrain it, but challenges it to motivate it intellectually. He wants to be an artist, wants to participate, wants very much to see an engagement across disciplines, between artistic and architectural forms. It seemed to be a pet project for him. Despite the square footage, he and Eileen know that they've managed to build the house completely in scale with the landscape. It's nested into the site.

In their former house, every piece of furniture was an object; everything was an art piece. But there was too much furniture, architecture, and art. The proximity between things was too close and didn't tolerate perspective; everything was completely compacted. They wanted their new house to be a museum, but a museum for living, with more breathing room between objects, spaces, and artifacts. Calvin Tsao has been tying the project together, and Monica and I have been doing the built-ins in their offices; we're playing a more deferential role, playing with surface and texture.

From an aesthetic point of view, there is definitely a difference between him and her, a different set of priorities; that's why their spaces are divided into "his and hers." The furniture is molded to fit their lives' respective styles—for him, carbon-fiber furniture, and for her, plywood.

However, it's not just about them, but about how you design things that resonate with the overall architectural ideas—for example, this spiraling object that's at once transparent and toroidal. It's a courtyard building, horizontally voided, but also vertically voided. When you're entering the living room, you're looking at the lake but also up at the sky. Crudely speaking, it's a free-form house. It doesn't have the spatial ordering of cubic volumes of space. It's characterized by figurative moves. You literally see a snake coiling around itself, the ceiling billowing, and it has a really sculptural presence. We've responded to that. Much of our role was how to negotiate between the clients and the particularities of the house. In order to create a backdrop, we needed to graft our furniture onto the lineaments of the geometry, and the sensibilities of the spaces. Some of the furniture, the plywood, was built in slats that conformed to the body of the space, but once they are free of the walls, they're liberated as an evolving, moving, sculptural object. In his office, the vertical support for the shelving system is concealed, and the dynamism of the house plays itself out in the dynamism of the bookshelves. You don't want the furniture to overwhelm the larger project and the landscape that runs through the building. Something about the landscape here is at stake.

joseph giovannini Architecture Critic

Bob Davoli and Eileen McDonagh dared to build differently—to want a house with a "moving center," and to not want Howard Roark as the architect. Did it work?

Although the word Post-Structuralism seldom, if ever, surfaced during the many years of the design and construction of their house, Davoli and McDonagh effectively embarked on a project that embodied two aspects at the heart of Post-Structuralist discourse. They were the exceedingly rare Post-Structuralist customers.

Davoli did not want an architect who would import a signature that would give the project a predetermined vision and an overarching narrative. Instead, he wanted to initiate a collaborative relationship with the architect in a process that would become a path on the way to an unpredictable result. The vision would emerge almost organically from the collaboration, and the journey would be as important as the completed house. Davoli was willing to pay for, and support, the side trips on a non-linear excursion.

If Davoli and McDonagh did not want a linear journey, they became interested in the interior of the house itself as a form of journey. She eschewed a conventional layout with traditional axes and symmetries that would make the house predictable, but spoke of a design with a moving center that would yield a multi-perspectival experience within a site, to be seen from many shifting points of view. He spoke of the interior as an unfolding experience. With curves breaking with and against other curves, the house, an evolving experience, like rounding hillside bends on a hike in the wilderness.

Neither was interested in a straight shot, either temporal or spatial, to an end result, but rather, in an unfolding process that would yield a space of becoming. Both the process of designing and building the house—and, for the clients, the act of eventually living in it—would invite all members of the extended design and construction team into an interpretative participation in its making. Clients and occupants would generate meaning through their own personal lenses as they lived both the process and the architecture. There was no single script. Any truths were multiple and shifting.

As if on a road test of theory applied to reality, Davoli and McDonagh were living the enquiry Jacques Derrida famously launched when he questioned what such architectural terms as foundation and structure meant in philosophy. He said that if the terms could be challenged in philosophy, they could be challenged in architecture—that architecture was a test case for philosophy. He wrote about moving centers, basically challenging the singular "correct" interpretations in favor of multiple views and truths. Before him, Roland Barthes had written his seminal essay, "The Death of the Author," in which he advocated drawing a distinction between the literary work and its creator. Authorship by a single person imposes what Barthes considered "interpretative tyranny" on the work. Davoli and McDonagh were making a pre-emptive strike on the house by separating the architect from the design, and pluralizing the designers. No single architect would be the sole designer, the sole agent of meaning and vision.

The architect who would be the un-Roark, then, would arrive without the traditional mantel of the architect. Derrida, Barthes, Davoli, and McDonagh removed the mantel of authority. Or, as Martin Buber would say in a political and social context, the clients "opened a closed system." Biologists call it "hybrid vigor."

Far and near, then, there was much intellectual background to support what seemed a unique enterprise, and the clients chose Warren Schwartz, an architect who rather bravely stepped into a force field in which he would be just one of the agents rather than the single determining force. Schwartz, who has always been a proponent of experimental architecture (at one point collaborating with Gehry on a Boston project), had the curiosity to enter into a new kind of architect/client pact, and unlike most architects of his reputation, he was willing to play catcher to the client's pitcher. The design would not be produced from a pre-existing vision, and it would not be formalist, but would emerge from the process, almost by self-generation. Davoli and McDonagh, major contributors to the Boston Institute of Contemporary Art, are denizens in good standing of the Boston art world, and decided to turn the house into a project that would include other invited voices. They would ask artists to create installations integral to a larger, building-size host. The clients conceived the building as an art piece, and to a certain extent, turned the process into an unpredictable happening, or construction performance, without the set score that controls almost all serious construction. Metaphors from the art world help to explain a process with little precedent in architecture.

Although Davoli said he didn't want to ask Frank Gehry to architect the house— because, he felt, Gehry would simply do a Gehry—the Los Angeles architect created in his own Santa Monica house, built in 1989, perhaps the only example of a contemporary work that established a precedent for the new house in Lincoln, Massachusetts.

For many years during Gehry's long gestation as an innovative architect, he befriended Los Angeles artists. His great originality was to step outside the then current mind-set of the profession into the mind-set of the artist, dropping even the dependency on the drafting board, adopting a sculptural rather than planimetric understanding of design. Local artists at the time—Ed Moses, Tony Berlant, Billy Al Bengston, Ed Ruscha—were building their own lofts without plans, and they were creating installations within galleries and museums. When Gehry decided to build his own house, he bought a milquetoast Dutch Colonial bungalow and created numerous interventions within its fabric, inserting installations influenced or inspired by his artist buddies. Gehry reinterpreted their ideas in a new architectural context, with a result that is heterogeneous rather than homogenous. The house was punctuated by different architectural episodes of different paternity. The Dutch Colonial, insipid though it was, was the binder that held the anecdotes together in a loosely ordered composition.

The Davoli-McDonagh house is analogous. The couple invited a number of artists to design site-specific works within the project. Landscape architect Mikyoung Kim designed a poetically evocative field of nature around the house, with a concatenation of standing Cor-Ten steel vertebrae that forms an unusual paddock for the family dog. Ilan Averbach did a sculpture for the entrance. Architect Calvin Tsao did the furniture and chose colors. Office dA created built-in furniture in the home offices.

For most of the artists, however, the collaboration was a question of parallel play; their work did not change significantly because of any interchange with the architect. Perhaps Kim and Averbach worked in reaction to the context of the established design, but without significant exchange or intermingling with other artists. With his usual, perfect visual pitch, Tsao chose atmospheric colors and designed furniture that worked with the space, but he worked not unlike other interior designers—certainly at the direction of his clients, and in consideration of the architectural context. Office dA took the measurements of the commission, laminating "his & hers" work spaces with built-in furniture, reacting with their usual sensitivity to context. There may have been collaboration between the clients and the artists, but there was more respect than collaboration between the artists and the architects.

49 north facade

Even though collaboration became a mantra of the project, these cases amounted to an act of juxtaposition, or design at a smaller scale reactive to the larger structure. The major effort at collaboration really occurred between the clients and the architect, especially in areas of the house that Davoli wanted to revise, to escape "the box," as he frequently said. Specifically, he asked the architects to revisit the fireplaces and, the kitchen island; McDonagh asked for changes on the back porch. The architects also designed intricate, beautifully turning bamboo woodwork, and cliffs of slate in the master bathroom that take on a life of their own, independent of the house.

Clients often ask for changes in a design, but artistic considerations motivated many of these interventions. The result was that each of the changes became a "piece" within the larger environment, like an installation within a museum. The difference was that the pieces were site-specific and permanent.

The overall design of the house as a floor plan itself emerged from the clients' desire for softer forms that appeared to be gentle within the sensitive landscape, and capable of creating geometries that could generate and activate moving centers. The curvilinear forms, though they resonate with other contemporary work by prominent architects, had little precedent in Schwartz's opus. They evolved primarily from the wishes of the client.

While the Gehry and Davoli-McDonagh houses, then, are characterized by highly individualized, episodic "installations," the anecdotes are embedded within entirely different contexts. Gehry's pieces nearly overwhelm the Dutch Colonial at its core, especially the corrugated wall that wraps the ground floor, and the chain-link fence festooned in the upper reaches of the façades. The anecdotes structure the building, changing its inherent reality, as in Chopin, where grace notes seem to expand to become a decorative filigree that structures the music.

The house within which Schwartz embedded his anecdotes, however, is not as distinctive. The host house did not provide as much resistance to the interventions, which therefore, did not engender the same artistic tension. Schwartz, working outside his usual vocabulary, which is often angular, simply extrudes the curved walls from the curving plan, without interrupting the verticality. The complex plan yields a surprisingly simple, if not simplistic, expression, missing its own potential for complexity. Furthermore, the extrusions are built with a glass-and-aluminum, commercial curtain-wall system, and the house that would be art suffers from the commercial overtones the materials bring to the project.

The anecdotes, which were done piecemeal, do not achieve critical mass within the house, and they do carry to the outside. Their collective weight is insufficient to win their confrontation with the house in which they are embedded. While the interior, with all its turns, is experiential rather than simply picturesque, the relatively undisturbed façade remains problematic, bland in its materiality and surprisingly stiff in its three-dimensional geometry. The leading edge facing the drive is awkward and does not attain the elongated sweep Richard Serra builds into his curvilinear steel sculpture. The architectural concept suffers from the fact that the exterior and interior walls are curved only in plan, but otherwise are straight and uninterrupted. The exceptional piece that proves the rule is the powerful ceiling, which floats over the living spaces and the brow of the exterior like a cloud. It reads both inside and outside.

This is a brave experiment, but more successful in its nerve and ambition than in its result. By avoiding the Guggenheim syndrome, in which the building sometimes challenges the art, the collaborators did not create a Guggenheim that could itself be challenged. The process pre empted the authority of the author, leaving a weak case to test—a context that did not have the presence of Gehry's Dutch Colonial to resist, and test, the interventions, as counterpoint. Schwartz, who at the time of this writing is building a brilliant house for himself in western Massachusetts, achieved there a strong, straightforward design that is powerful all over. Working in his own idiom, the author has restored his authority, and though the project is less ambitious in its process, it stands as a strong work of architectural art. As the Abstract Expressionist painter Sam Francis once said, "The individual is always the carrier of the vision."

douglas belkin

Wall Street Journal Reporter

***Wall Street Journal* Reporter and Former Staff Reporter for the *Boston Globe*; Author of November 13, 2005 Globe article, "Despite a drop in the real estate market, a surge in $10 million houses is ushering in a new Gilded Age. No one is happier than local craftsmen."**

I was covering suburban life for the Boston Globe, and a lot of the life in the beat in Boston was class issues, money issues, culture issues, That's one of the things that stands out there, because there's so much money. I'd written a piece about Walden Pond the summer before, which is not far away, and I'd taken a flight over the Pond to get photographs. I'd spent a lot of years biking in that area, and I was amazed at the estates you could see from the air that I had no idea existed. And that prompted the question as to what's going on with these estates. I just assumed at first that they were old money, that the places had been there a long time, because that area has been developed over many years, hundreds of years, most of it right near the shot heard around the world, Lexington.

So I began to look into that, and I started to talk to builders, you know, in passing; whenever I'd come up on one, I'd ask him about the really high-end estates, and I think at the beginning I was curious about maintenance, just the aspect of maintaining them. And then someone turned me onto this company, Thoughtforms, and they were interesting because that's what they do. So the story sort of generated from there, that these extraordinary gilded mansions were being built now that superceded anything that had been built for a long time. Just the pace of them and the scale of them—that was the genesis of the story. The time between the flight and the story may have been a year and a half.

Thoughtforms directed me to this house; this was their baby, and they were happy to get some publicity. There was a clip about Davoli, and I read a little about him. He's an amazing guy. He's self-made, he didn't come from money, he made it, and he's obviously brilliant at what he does. I called him up, and it took a while to get a hold of him, busy. He's really proud of it. I mean, it's an amazing house. I don't think I met him, I think I just spoke with him over the phone, one interview for just fifteen or twenty minutes. My exposure to him was not substantial. I was out there two or three times, got a tour, and it was amazing to walk around. What was cool was the guys who were building the house were there, and I spoke to them, and the foreman in particular, quoted in my story, was a guy from South Bay who was a real hoot, with a heavy South Bay accent. Had a wicked sense of humor. He told me he had a real knack for trig and calculus, and he was good at figuring out these hyper-individualized rooms. He did a lot of calculations to get everything right. He'd been on the house for four years. He was the GC's main guy. He was good, because he explained things in plain English, how things were put together.

I'd never seen anything like it. You know, I'd been to Newport that summer, and I go up there every summer and I go look at houses there. I do enjoy that stuff. I'm just fascinated. I don't have any architectural training, and I'm not particularly

50 stairway to Eileen's office - third level

aesthetically inclined. I'm a layman. But it's all just impressive. So the house was not finished, but the staircase blew me away. The house felt cold to me, not particularly warm, but it was impressive. It was stuff I'd never seen. The amount of care that went into everything was mind-boggling.

So I became aware that there were these mansions set way back that I really didn't know existed. In the story, I mention a few other homes. And I'd been to mansions in the North Shore, the Beverly area, like Beverly Farms, that are cool. There's a lot of old-money homes there, sort of castles, with these pieces of land on the water that you can't make up. They're all crazy good. So I was in and out of a bunch of them, but the Davoli house was definitely the most impressive. There were two homes built, I heard, for $30 million in Weston, but I just didn't have the time to get there. I had to feed the beast at the same time.

What struck me at Davoli's was that something as seemly insignificant as the place where you hang up your toilet paper could cost that much money. It almost struck me that you had to scour the world to find a way to spend, I don't know, six, ten, twelve, fifteen thousand dollars on something that you can buy for eighty-nine cents that's obviously as effective. It obviously doesn't have the same pedigree, but it doesn't cost as much. My perspective about this piece is I'm writing stories for my mom, the guys I play softball or racquetball with—you know, people who don't live in this world and haven't been exposed to it. Same way I am. I think the reason I picked out that detail to write about was because it was the most extreme example of how expensive something could be if you set your mind to it. You know, a $6,000 pizza oven kind of makes sense if you want to make pizza. That's what I recall the pizza oven was, something like six grand. I don't know; maybe I'm off. And I think in another house, a guy spent $600,000 for the plates for outlets, which I had to check several times because it seemed so preposterous, and my editor insisted that I check it again, because he said it was not possible. And that's who I was writing the story for: people who are not architects or architecture critics but just regular folks who live in the area who would be surprised by the capacity to expend money on things like that.

I loved the view, and I loved the trees. After college, I spent a summer living in a tent in Wyoming, and my view was just as good from that twenty-dollar tent. But the view is beautiful. They hadn't finished the house, but they were saying cool, Star Trek-ky stuff, space-age stuff. Like, if you were going to leave your bedroom and walk to the kitchen down the hall, or whatever, then there was a way that the floorboard lights would light up, sort of ahead of you, and it seemed neat, and also pragmatic, which appealed to me. The house itself is so non-pragmatic, so abstract, that I had a hard time getting my head around it.

Compared to other mansions, there's nothing like it. Most others were building versions of Colonials, and there were elements of things I recognized in every other house. That was not the case with this house—there were very few things I recognized. I suppose it sort of jars you, and maybe that's what he was going after. It made you think about the house, that's for sure. You didn't take it for granted, because it was so different. The other homes I saw—I understood them more, and maybe for that reason, I didn't look at them as closely. This is the one that made the deepest impression. I saw at least two more, maybe three, but nothing as grand or impressive. In Weston, I spoke to the building inspector there, and he said that those $30 million mansions were conventional to the point of boring. And I just ran out of time.

I remember asking Davoli if there was an analogy to the robber-barons' homes of the Gilded Age, and he took offense at that. He said that they didn't do anything really interesting with their architecture. That they were just big, grand homes, that I'm trying to do something more interesting, much more compelling. So I think he was making a statement, leaving a legacy, doing something really original, breaking some rules and breaking ground. I remember thinking that I thought this house might be appreciated more in a hundred years. It was a little too jarring. Not to sound too crass about the part about the money, but class and cash run so much underneath motivations and perceptions, especially in that part of the world. Certainly, people who are reading have a hard time just getting around the amount of money expended. I'd been covering crime in the city for three years. I'd spent a bunch of time around shootings, in Dorchester and Roxbury. There's a huge amount of need there, and those places are a jarring juxtaposition against the expense of the Davoli home. The recording of my days as a cop reporter in the tougher areas of Boston, with the shootings, was a background track running in my head. How could you not put a price tag on the house?

I was living in a 700-square-foot apartment, and the Davoli place seemed so bizarre to me. But one of the cool things about being a reporter is that you get to step in and out of different worlds. So I've been to the poorest, most violent sections of Boston, and I've been to the wealthiest corners of the greater Boston area—and yeah, sometimes they collide. And of course the vast majority of folks are squeezed in the middle, and class envy and who's got what and keeping up with the Joneses, all that creates a lot of friction and spins off a lot of social stories in the Boston area. And not just in America, but in the world. Class envy. As a reporter, you're looking for sources of friction and tension, and you don't have to scratch too deep to find that, and certainly, it's easy if you're writing about something that costs $16 million. I don't know how many thousands weare spent for a stainless-steel toilet made for prisons.

There is so much money in the country now, and it's generating more money. I'm at the Wall Street Journal now, so we cover people with crazy money. In Boston, it's financial money. There are a couple of billionaires in town, and they've spun out a few other hyper-rich guys. But these hedge-fund guys who work with the big numbers, they're unto themselves. Davoli was kind of extraordinary because I think he was a local kid. I think he went to school here. I think he was a guitar teacher at some point. There's a good story on the cover of Fortune, or maybe it was Forbes magazine, from the late nineties or early 2000. You can pull it up—that will tell you everything. It's a good financial profile. He put together a couple of big-time, high-risk deals that paid off in multiples. That launched him. More power to him. I can respect that.

The most striking thing about the difference between the old robber-baron mansions and his was that there wasn't a big lot. You expect a house that cost $15 million to have a big chunk of property and privacy. But it was at the end of the road, and as I recall, at the beginning of the road, there was a car on concrete blocks. You know, the road wasn't finished, and I knew that they were going to put down Belgian cobblestones and get the driveway finished. But you could throw a baseball to the next guy's house. Those guys in Newport—you needed a rocket launcher there to get to their neighbor. That's not true of all of them, but they have those vistas over the ocean, so you feel you're in a grand place. This was certainly peaceful, but it didn't have that majestic sense to it. It had that high-weird sense to it. Kind of cool.

From the little I spoke to him, and from the tradesmen's impression, I would say Davoli wasn't trying to impress. He didn't strike me as aloof. He was about doing something original and different, and he just happened to have the bucks to do it.

But you know the focus that made him wealthy in the first place, what probably made him a great financial guy, that's also probably what propelled this house into the stratosphere in terms of detail and pricing. I guess. It's the personality type that gets obsessed. I bet it connects.

I was impressed by the house, bowled over by it, moved even. Did I like it? I don't know if I was comfortable. Remember that it was also under construction, there were nails, things half-done. Did I like the house? I don't think it's how I'd spend $16 million if I had it, but I wouldn't mind waking up on that lake. I guess I have a complex set of reactions. I don't know that I liked the house, but I was impressed. With the force of energy from all those people, those 400 people working for four years, it's hard to separate the work from the result. From an aesthetic point of view, I didn't like the house. It was okay. The inside was better, but they hadn't landscaped. There was a fence on the outside that they were working on. I vaguely remember it, but I remember thinking, "Ehh?"

My article ran nationally. It was a New York Times newswire and it was picked up, so from where I'm sitting, the feedback was that other editors thought that it was an interesting news story. And that was good. It was cut up into a 500-word piece and put in the front of the Metro, which is a little freebie that the Globe now owns that people read on the subway. I never saw that. It ran in the Chicago Tribune, and I don't know where else.

The one thing I'm afraid of is sounding mean-spirited, and that is not what I want to convey here. From where I'm sitting, the fact that this guy could do this, that he made the money—God bless him. And when I reread the story, I thought that I did a pretty good job with that. There's an intense amount of money that goes in, and that can be off-putting to some people. On the other hand, this guy's making a statement, he earned it himself, he put a lot of people to work, and there's a huge social benefit to that. I thought I balanced that. I could understand why the Davolis would be annoyed, why Davoli wouldn't like the story, but I think that when I reread it, I thought it was fair stuff. But I'd be sad if I came across as a scold or mean-spirited. It's not how I feel. Again, more power to the guy. I don't have the eye or the training to judge the building. But you know, I'm probably going to be happy with a really good burger, and not appreciate a Kobe beefsteak for $110. You know what I mean, if you don't know so much. That's kind of how I feel about architecture. It's just what hits me, I'm not a student of that stuff.

51 main entry from south

52 garage ramp

53 balcony off master bedroom

Eileem McDonagh, *continued from page 61*

As far as I can tell, this has been a project with no adversarial principle. Even when we disagreed, it wasn't about who would win. The argument was what served the idea best. It was never that I won and Warren lost, but only that he couldn't see what I could see. Had he pressed an issue, I would have gone along, because he's the architect. Any adversarial principle came from Lincoln and from outside people, and that's good, because, again, that's life. But as far as people working on the house, there were no breakups. People grew closer, and people have become friends. I don't know how long the friendships will continue after the house, but some will. Because everybody who works on the house is goal-oriented. They're people who don't like to fight, but create, and their creativity is not a zero-sum game. It's all additive.

Community can be a limiting factor in people's lives. It sets up norms, and can be a conservative experience. But a community that allows for individuality and difference is what makes a community grow. This project is a community of people who are all distinct individuals with great abilities distinct from everyone else's. But we are working together, producing something that's larger than the sum of the contributions. Without any one of us, it would not be as good. Without the house, the landscaping would not make sense, for example. It's all reinforcing the individuality that everyone contributes.

I hate the great hero myth—the great writer, the great artist—but I do think creativity is defined in terms of the individual. Some of the opposition that came from the neighbors, and from a writer who writes for the Lincoln Journal, and letters to the editor—and there was an article in the Boston Globe—didn't focus on this house, but a category of houses. How do you justify the implications, the ethics of building a house like this defined in terms of the expenditure of resources? Should a house like this ever be built or should those resources be used in other ways? A house like this raises these questions, but I think these are questions that must be considered in every aspect of life.

In terms of resources to create something like this, especially if you think of what could have been done instead— like helping people in need of food, not only here but all over the world—it's crucial to consider, for example, how many houses you could build for the homeless instead of this house. However, the question—or problem—of allocation of resources goes beyond the house. It is at the heart of all expenditures that are made in lieu of contributing to the basic needs of humanity. For example, why, (as some people ask,) should the American state spend money exploring space when there are people at home and around the world, including children, starving to death?

To me, the answer is difficult, but I do make a distinction between consumption and creativity. This house does not represent consumption in the sense of using up resources without there being any lasting contribution to others. This house is not like buying something already made or something for which there is only a comfort rationale, like buying a jet airplane for travel or buying expensive jewelry distinguished primarily by the size of its stones rather than by artistic design or craft. To the contrary, the rationale motivating this house is creativity, both individually and collectively. This house does not represent the consumption of resources but the use of resources to fund a wide array of artists who were basically given a blank check to do what they wanted. And comfort was barely a consideration, much less resale value, true to the artistic mode of the enterprise. It is crucial to allocate resources to humanitarian needs, as individuals, as a society, and as a political system, as well as to creative projects. And to me, it is not so much a question of either/or as it is a question of balance. As individuals, Bob and I do try to achieve balance in the way we allocate our resources to include the arts as well as attention to the social and economic needs of others.

To me, therefore, the ethical issue centers on the balance between allocation for creativity and allocation for humanitarian needs. I do not think it would be ethical to take all funding for space and use it for meeting utilitarian needs, because I think human beings also have creative needs that should be fostered for individuals, society, and the government. These creative needs are vital for individuals and communities alike. It is as if there are capacities of the human mind that need to be explored. Creativity is like a food drive. You can't survive completely without it.

So, that's how I justify some activities, like building this house. It's part of a definition of what a human being is. I do believe that creativity and curiosity are human needs, and that they should find constructive human expression. You could say they hurt the world passively by taking resources away, using the resource of people at a subsistence level. But that would give up this other need.

Everybody I know on the project has taken personal pride in how they've solved a particular problem within their sphere in the house. For example, the Cor-Ten fence that Mikyoung did—it's not that she had to do this; it's that she wanted to do it. I like this idea of people having and taking the opportunity to be creative. And that's a need different from accumulating things—consumption— to attain status or power. I would say that this house has nothing to do with power. It demonstrates experiments, and the hope that they work. There are a lot of new things that no one working on the house did before, like the fence. Or like the dining table, by Calvin and Zack, which they designed as a connected series of vertebrae. It took some doing to get the shapes right. The fireplaces are very good, too. I do believe creativity is really a need.

So, I believe the house stands for the creativity of individuals working within a supportive community, or, as William James put it, as inscribed on the building at Harvard bearing his name: "Without the community, the individual dies, but without the individual, the community stagnates."

As for the articles, the really important thing is that people spoke, and that they felt free to speak. We never built the house with the idea that we were expecting social approval. Everyone we know whom we consider friends likes the house, so far. That doesn't mean that everyone will like it. We didn't build it with that expectation. We saw it as following in the Gropius tradition of Lincoln, and it turns out some people don't see it in that tradition, and that's their opinion. I'm glad they expressed it. I was idealistic about what Lincoln stood for in terms of the value of modern architecture and conservation, and the common points of reference meant different things to different people.

But for me, this house makes me feel at home because it captures the feeling of walking in the wilderness. Because of the depth of the house, it just seems there's nothing there but the trees and the water beyond, and all the changes the days and seasons bring. Most structures can be comfortable, but they're a static environment. This house doesn't feel static. It's the combination of the curves, the indoor/outdoor vistas, all depending on where you're standing. Even if you move two feet, you'll see something different. It's like a hike, along the contours of a cliff, with the revolving angles of the view. And it's especially like hiking on the East Coast, where there's so much water and so many mountains and you're in foliage all the time, with ferns, huge bushes, and trees. Not like the West in the Sierras, where at about 6,000 feet, you get extraordinary vistas of distance, perspectives in depth.

54 view east from eileen's overlook

55 north facade at sunset

I feel that there's a dynamic to the experience in the house because it's multi-centered and depends on your point of view. It gives you a range of perspectives that free you up rather than fix you to a single spot as you walk into the house. As opposed to fixed, centered viewpoints, it's a combination of diversity and integration. Not that it fractures views. There's a set of distinct views, and you feel comfortable and centered with each of them.

In most houses you walk into, there is a center, and then you go right or left. But here, you can't walk in and stand in the center of the house. It's more of a moving center that yields different viewpoints, depending on where you're standing. That's what you see, and depending on that experience, that's what you believe about the house and the site. There is no one center, and no one belief system. People find their own place it the house, and draw their own interpretation. It's not imposed on them by geometry or organization. It's not as though this is the study and only the study, and likewise, for the dining area, which could only be the dining room and will always be the dining room. Things shift; they change. The major principle was to minimize and even obscure the boundary between outside and inside, and the way we did that was by cultivating many perspectives. It's a series of unfixed, mysterious, floating viewpoints.

Initially, Warren's design called for a stairway and an elevator just at the entrance, which would be customary. But I remember asserting myself, saying, "No, we can't have a stairway and an elevator in the center, because we need to walk in and feel we're heading straight out through to the pond. When we step in, we want nature to continue right through the house." And it turned out to be an argument, because they'd already designed it. The only principle I'd ever felt strongly about was obscuring the boundaries, having it as close to being a house as you could without building a house. In fact, the way the house is, with all the glass, it's as close as you could come to being there without a house on the lot.

Even though the stairway was built, we made it into an art object itself, a gang-plank, as though up to a boat, but built in glass. So it's dynamic, not static. And there's another glass stairway like it up to the second floor. We moved the elevator somewhere else, establishing a shaft up to the second floor. As it turned out, Warren is really happy with that decision. Everyone's happy with it.

It's only up in my study, the right side of the house that is really closed off, where you don't get the interplay of outside and inside. And that's the area where I'll try to write. When writing, I don't want to be distracted. I write facing the wall, otherwise I can't write.

I'm pleased that the house seems to me very accessible and friendly and not at all prepossessing. People feel comfortable. I think that's because of the curves. You don't get the feeling of the space overwhelming you all at once; you can find protected areas. There's some demarcation of space, but not rigid dividers. You can stand in little sections and yet see beyond, and people feel comfortable. You're cozy with people you're talking to, but not isolated from people. It does accommodate large numbers, certainly over a hundred. We've had birthday parties for Bob, and a halfway house party. We've had fundraisers, and we've had two weddings, one for Bob's guitar teacher.

I believe that the house belongs to everyone who has been part of it. I own it only to the degree I participate in its life, like my dog. You don't own a thing if it's alive. The house belongs so much to everyone who contributed to it. I feel it when I walk in, I see them. They'll always be welcome. We look forward to sharing life with them. The house is as close to a living thing as a house could ever be because of the community that's been brought together to build it. I do like that feeling, of something that's ongoing and alive.

56 screen porch - master bedroom above

"...I'll sink down and nestle into the pond."
— Bob Davoli, owner

credits

Project Name

Davoli-McDonagh Residence

Owners

Robert Davoli, Eileen McDonagh

Location

Lincoln, Massachusetts

Architect

Schwartz/Silver Architects, Inc.

75 Kneeland Street, Boston, MA 02111

617 542-6650

www.schwartzsilver.com

Design Team

Warren Schwartz, Robert Silver *(principals)*; Michael Price, Steven Gerrard *(project architects)*; Michele Baldock, Jonathan Bolch, Philip Chen, Sam Choi, Christopher Ingersoll, Richard Lee, Sandra Saccone, Paul Stanbridge *(project team).*

Landscape Architect

Mikyoung Kim Design

Interiors

Schwartz/Silver Architects; Tsao & McKown; Office dA

Engineers

LeMessurier Consultants *(structural)*; Sun Engineering *(hvac)*; Johnson Engineering and Design *(electrical)*; McCarthy Mechanical *(plumbing);* Nitsch Engineering *(civil).*

Consultants

Stephen Stimson Associates *(landscape/site planning)*; Dewhurst Macfarlane and Partners, Robson & Son Engineering *(structural)*; Simpson Gumpertz & Heger *(building envelope)*; Walter Moberg Design *(fireplace)*; Berg-Howland Associates *(lighting)*.

For additional consultant information, please contact Schwartz/Silver Architects.

General Contractor

Thoughtforms Corporation

Photographers

Alan Karchmer, Sandra Benedum *(Photo Stylist)*, Shellburne Thurber

Schwartz/Silver Architects

Date of Design

1999-2006

Date of Construction

2001-2006

book credits

Book Concept
Oscar Riera Ojeda

Project Coordination
Kat Monaghan

Copy Editing
Nirmala Nataraj

Graphic Design
Linda Prescott

Project Assistance
Robert Silver, Michael Price, Penn Ruderman

Foreign Editions Sales
Gordon Goff

Production
Oscar Reira Ojeda

Color Separation and Printing
ORO editions HK

Case
Saifu custom died in Japan from Toyo cloth
with dutch boards

End paper sheets
140 gsm wood-free
from NPI, Tokyo

Text
157 gsm Japanese
White A matt art paper.
An off-line gloss spot varnish was applied
to all photographs

copyright

ORO *editions*

Publishers of Architecture, Art, and Design
Gordon Goff & Oscar Riera Ojeda – Publishers

West Coast:
PO Box 150338, San Rafael, CA 94915

East Coast:
143 South Second Street, Ste. 208, Philadelphia,
PA 19106

www.oroeditions.com
info@oroeditions.com

Printed in China by ORO *editions* HK
ISBN 978-0-9774672-9-7

Distributors

North America
Distributed Art Publishers, Inc.
155 Sixth Avenue, Second Floor
New York, NY 10013
USA

Europe
Art Books International
The Blackfriars Foundry, Unit 200
156 Blackfriars Road, SEI 8EN
United Kingdom

Asia
Page One Publishing Private Ltd.
20 Kaki Bukit View
Kaki Bukit Techpark II, 415967
Singapore